Diffuse Pollution

An introduction to the problems and solutions

Neil Campbell, Brian D'Arcy, Alan Frost,
Vladimir Novotny and Anne Sansom

Published by IWA Publishing, Alliance House, 12 Caxton Street, London SW1H 0QS, UK
Telephone: +44 (0) 20 7654 5500; Fax: +44 (0) 20 7654 5555; Email: publications@iwap.co.uk
Web: **www.iwapublishing.com**

First published 2004

Printed by TJ International (Ltd), Padstow, Cornwall, UK
Copy-edited and typeset by HWA Text and Data Management, Tunbridge Wells, UK

Photographs by Brian J. D'Arcy unless otherwise indicated.

British Library Cataloguing in Publication Data
A CIP catalogue record for this book is available from the British Library

Library of Congress Cataloging in Publication Data
A catalog record for this book is available from the Library of Congress

ISBN: 1 900222 53 1

Contents

Preface

This book is the final output of a project undertaken under the auspices of the Specialist Group on Diffuse Pollution of the International Water Association (IWA). The first output was the diffuse pollution video, *Nature's Way*. Launched in 1996, *Nature's Way* was the first video project of the then International Association on Water Quality (IAWQ) – one of the predecessor organizations that merged to form the IWA. The video was well received around the world, including being translated into Japanese. The project was initiated when Vladimir Novotny was Chair of the IWA Diffuse Pollution Specialist Group, and the book of the video (this publication) was carried forward when Vladimir Chour took over the chairmanship of the group. The project has grown ambitiously over the past few years, although the simple intention of being an introduction to diffuse pollution and the best management practice approach to controlling it has endured. It was realized that the socio-economic factors that are essential in the delivery and implementation of control measures must be considered too. Finally, diffuse pollution is a cross-media issue, and although the focus of this book and the IWA group is water, the origins of the pollution include atmospheric deposition, land-use change and runoff from land. Diffuse pollution is often incidental to land-use decisions; a corollary to normal economic activity. Diffuse pollution is often the environmental consequence of that economic activity; typically not understood by the 'polluting' sector, because no obvious polluting effluent is produced and the pollution arises because 'that's the way the land is farmed /roads are built and drained' etc. The logical concluding chapter is therefore a look at diffuse pollution and sustainability.

The authors are very grateful to the steering group for this project who have helped over the years, either by contributing material, drawing important work to the attention of the authors, or reviewing chapters or sections of chapters. Each chapter has been reviewed by at least one, more usually several, different professionals including scientists, engineers and regulators. Of the authors, Vladimir Novotny and Brian D'Arcy was/is respectively founder chairman and current secretary of the IWA Specialist Group on Diffuse Pollution. Steering group members Giuseppe Bendoricchio, Nita Tonmanee and Ralph Heath were or are committee members of the group. Bryan Ellis, Larry Roesner and Shoici Fujita are also active members of the IWA/IAHR Urban Drainage Specialist Group. The International Water Association offers unparalleled opportunities for professionals to meet and help each other, and develop their interests, this book being just one of many collaborative projects under the auspices of the association.

This book aims to be useful for students and for those academics and professionals to whom diffuse pollution is a new subject. Its value is greatly enhanced by the BMPs performance datasheets in Appendix 1, and the authors are grateful to John Hilton and the Centre for Ecology and Hydrology (CEH) for the rural BMPs information, and Ben Urbonas and the American Society of Civil Engineers (ASCE) for corresponding urban BMPs performance data. Many of the projects referred to in the text are on-going and websites will provide easy access to follow their progress in subsequent years. A great deal of useful information is available via the internet of course, and the authors are grateful to Morag Garden (of Scottish Water, and treasurer, IWA Diffuse Pollution Specialist Group) and Peter Wright of Scottish Environment Protection Agency (SEPA) for compiling the list of websites that is also included in the appendices. The information in the text is enhanced by use of photographs throughout, mostly taken by the authors, but we are grateful to Dov Weitman of the US EPA for supplying additional material (as well as a lot of useful US publications and contacts).

Due acknowledgement is also made to the following who have variously helped in commenting on parts of chapters, carrying out literature searches, obtaining esoteric publications, and providing figures and example case studies for the book: Andy Griffiths, Evan Williams, Neil McLean, Ken Pugh, Alison Dick, Peter Wright, Alastair McNeil, Jennifer Davidson, Tom Wild, Craig Wilson, Robin Clarke, Kate MacCalman, Jackie Vale, Phil Chatfield, Helen Richardson, Dave Griffiths, Kirsteen Macdonald, Tom Schueler, Tim Darlow, Morag Garden, Bill Parr, Eric Armet, Bob Bray, Paul Harrison, Kate Hart, Peter Johns, Stuart Henderson, Ken Iwugo, Bill Parr, Tilla Larssen, Peter Stahre, Torsten Rosenquist, Paul Younger, Manda Hinsch, Douglas Alexander, Michael Dunn, Alan Peterson and Vanessa Humphries. Thanks are also due to two key supporters of the initial *Nature's Way* diffuse pollution video project, without which this publication would not exist: John Tyson and Tony Milburn.

As usual, the opinions expressed herein are those of the authors, not necessarily their employers. In conclusion, although each of the authors has helped edit the others' contributions, as well being lead contributor on at least one chapter, the final edit has been mine, so all credit to the whole of the authorship team for any parts of the book that are useful, but the blame for any final changes to text rests here.

Brian J D'Arcy,
(on behalf of Neil Campbell, Alan Frost,
Vladimir Novotny and Anne Sansom)

The Authors

N.S. CAMPBELL

Scottish Director and principal engineer, Sir Frederic Snow & Partners, Edinburgh, UK. Co-author of SUDS design manuals for Scotland, and for England and Wales, and technical papers. Specialist advisor to several local authorities in Scotland on SUDS and flooding issues.

B.J. D'ARCY

Diffuse pollution project manager, Scottish Environment Protection Agency, Perth UK. Secretary of the Diffuse Pollution Specialist Group of the International Water Association. Co-editor/author of diffuse pollution books and papers, plus magazine articles and videos.

C.A. FROST

Environmental consultant with Soil & Water, Scotland. Formerly soil scientist with Scottish Agricultural College, SAC, and editor-organizer of three diffuse pollution and agriculture conferences. Author/co-author of various scientific papers and technical guidance documents.

V. NOVOTNY

CDM Chair Professor at Northeastern University, Boston, Massachusetts. Founder chair of the IWA Diffuse Pollution Specialist Group, author/ co-author of definitive diffuse pollution text books and technical papers.

A.L. SANSOM

Retired. Formerly ran the Environment Agency's Rural land Use Project, based in Leeds, UK. Studied agriculture and diffuse pollution issues in USA with a Nuffield Farming Scholarship and Yorkshire Agricultural Society Award.

The Steering Group

This project was initiated by the Diffuse Pollution Specialist Group of the International Water Association, represented at the outset by Brian D'Arcy and Professor Vladimir Novotny. The following have supplied information and advice for the project over the past three years and their support and expertise is gratefully acknowledged:

Prof Giuseppe Bendoricchio, Italy
Mr Arno van Breemen, the Netherlands
Prof JB Ellis, UK
Dr Phil Haygarth, UK
Dr Ralph Heath, South Africa
Dr Chris Jeffries, UK
Prof Paul Jowitt, UK
Prof Chris Pratt, UK
Mr David Riggle, USA/UK
Prof Larry Roesner, USA
Dr Kirsty Sherlock, UK
Prof Paul Younger, UK
Prof Shoichi Fujita, Japan

Useful websites

The following websites often provide diffuse pollution information and/or updates on published reports.

www.nal.usda.gov.wqic The Water Quality Information Center for the National Agricultural Library (NAL) is part of the U.S. Department of Agriculture's (USDA) Agricultural Research Service. The center was established in 1990 to support USDA's coordinated plan to address water quality concerns. As a focal point of NAL's water quality efforts, the center collects, organizes and communicates the scientific findings, educational methodologies and public policy issues related to water quality and agriculture. The center's activities involve three areas: communications, library resources and special projects.
www.environmentdaily.com Europe's environmental news service
www.newscientist.com Claims to be the World's No 1 Science and Technology News Service.
www.wef.org Water Environment Federation Home Page.
www.endsreport.com ENDS Website – provides up to date information on the environment.
www.iwahq.org.uk Home page of the International Water Association. General information on the Association and its activities and full IWA conference information are available at this site.
www.watercouncil.org.uk Home page of the international water policy think tank.
www.alphagalileo.org Home page of the internet press centre for European science and arts.
www.planetark.org/dailynewsstory.cfm/newsid the home page of Planet Ark – world environment news. Part of Reuters News Service.
www.epa.gov/OWOW/info/NewsNotes/ Nonpoint Source News-Notes is an occasional bulletin dealing with the condition of the water-related environment, the control of the nonpoint sources of water pollution (NPS), and the ecosystem-driven management and restoration of watersheds in the United States. It is produced by the Terrence Institute under an agreement with the U.S Environmental Protection Agency. It is distributed free of cost and is available on the Internet.

www.npwa.freeserve.co.uk/H20.html The UK based National Pure Water Association has recently posted this comprehensive, user friendly Water Research web page.
www.asahi-net.or.jp/~gp7y-mti/ In "Tokyo Water Net" you can read about Tokyo Metropolitan Waterworks Bureau and other articles concerning water issues in Tokyo and around.
www.ukwir.org. UK WIR facilitates and manages collaborative research for UK water operators. The UKWIR programme generates sounds science for sound regulation and sound practice.
www.water.org.uk Water UK represents the water and wastewater services suppliers at national and European level.
www.nwp.nl The Netherlands Water Partnership (NWP) is an independent body set up jointly by the Dutch private and public water sector to act as an international focal point for the exchange of information related to water and water activities.
www.europe.unep.net The European portal aims to provide on-line authoritative information on the environment situation throughout the pan-European geographic region, while highlighting key issues in specific problematic areas.
www.edie.net The online community for water, waste and environment professionals.
www.europa.eu.int/comm/environment/ippc/eper/ Since 1993 as a follow up of the UN Conference on Environment and Development (UNCED) in Rio de Janeiro (1992) the OECD has been encouraging national governments to establish PRTRs and providing guidance in their implementation. A PRTR, or Pollution Release and Transfer Register is defined as a national environmental database of harmful releases to air, water, land and waste. The database contains information on releases (emissions data) of polluting substances, reported annually by individual facilities. However, it may also contain information on releases from sources other than large industrial establishments.
www.ukadapt.org.uk Agricultural Diffuse Aquatic Pollution Toolkit. UK-ADAPT is a resource for researchers and funders to make everyone aware of projects that contribute to our understanding of managing catchments to decrease diffuse pollution.
www.smurf-project.info/ SMURF (Sustainable Management of Urban Rivers and Floodplains) is a three year partnership that started in August 2002. SMURF is supported financially by the EU LIFE-Environment programme. SMURF aims to tackle these environment problems on the Tame by integrating the planning and management if land-use, water quality, ecology and flooding. The methods developed by the SMURF project will be used as a model for work on similar rivers throughout the UK and the European Community.
www.environment-agency.gov.uk/youren/353880/362533/361306/?version=1&lang=e Tackling pollution upstream to bring Bournemouth's bathing water up to scratch. The partnership, which has been set up by the Agency in association with Bornemouth and Poole Councils and a number of local business and conservation organisations, has set up projects to decrease pollution into the Bourne Stream. In turn, this should have a favourable impact on the bathing waters into which it flows.
www.odenseprb.fyns-amt.dk Danish Odense Fjord Pilot River Basin work related to the Water Framework Directive.
www.dorset.ceh.ac.uk/River_Ecology/River_Systems/Diffuse_Pollution.htm Centre for Ecology and hydrology, UK based organisation; includes websites listing for diffuse pollution, and lists of BMPs with indications of performance and key characteristics.
www.sepa.org.uk Scottish Environment Protection Agency website includes diffuse pollution (see Initiatives in menu), bathing waters, SUDS, and relevant publications.
www.eea.eu.int European Environment Agency, based in Copenhagen, Denmark, publishes environmental assessment reports (for example Nutrients in European Ecosystems, 1999).
www.defra.gov.uk UK government funded research includes rural land use and diffuse pollution, including reviews of effectiveness of pollution control measures.
www.adas.co.uk UK agricultural advisory service site includes information on NVZs, organic farming and good practice codes, as well as links to the UK Adapt research listing.
www.relu.ac.uk The rural economy and land use site introduces the projects of the UK research councils in this area, including diffuse pollution and land use.

1

The nature and importance of diffuse pollution

1.1 INTRODUCTION

Diffuse pollution is the unfinished business of water pollution control that was recognized as a problem relatively recently, in the 1970s. For many years technology, business practice and regulatory activity have been developing to achieve ever tighter control of major effluent discharges such as industrial process effluents and municipal sewage discharges. The building of sewers and treatment plants in developed countries has been continuing for more than one hundred years and achieved remarkable successes. Today, in many countries, diffuse pollution is now the biggest remaining problem. It should not be thought that it is a new problem, however, rather its impacts were formerly masked by gross pollution from the major point sources noted above (sewage and industrial effluents). The impacts of diffuse pollution in the USA and Europe are not the same as the effects historically associated with, for example, poorly treated sewage where high organic loadings would cause low oxygen concentrations or toxic levels of ammonia to occur in receiving waters. Many industrial effluents would exert similar effects and cause the direct toxicity associated with the presence of polluting chemicals.

Diffuse pollution impacts are typically different in kind, often associated with nutrient enrichment, contamination of sediments, and with siltation of recreational waters and fish spawning riffles in rivers. These impacts can be very significant, especially for water utilities where expenditure of millions of pounds has been documented in the UK, for example (in D'Arcy *et al.,* 2000). Toxic impacts of diffuse pollution can be intermittent and difficult to identify with certainty. Rural examples include insecticide applications that involve directly spraying small watercourses in the headwaters of an arable catchment, or pollution due to contaminated run-off from sprayed fields. In urban catchments there is very often little life in small urban streams, which is a result of the combined effects of contamination by oil and toxic metals associated with vehicle traffic, plus municipal pesticide applications, dumping of chemicals and run-off from industrial yards. Because of diffuse pollution, very few urban streams, even without notable point source pollution, are unpolluted, although during dry weather (when diffuse pollutants are not mobilized) the stream flow can appear clear and clean.

Diffuse pollution is far from being just an issue in developed countries. In many parts of the world soil erosion is a threat to the sustainability of agriculture and the dependent human population, as well as a cause of water pollution. River and groundwater contamination by faecal pathogens associated with agriculture and dispersed settlements are the principal threats to public health. Chronic contamination by persistent pesticides manufactured in Europe and the USA (but exported since being banned in the source countries) is a continuing issue.

Gross impacts can still occur as a result of diffuse pollution, especially in developing countries where sewerage infrastructure is often absent and multiple, individually minor sewage discharges and seepages from squatter camps, sprawling townships and *ad hoc* settlements collectively exert major impacts on local watercourses and groundwaters. To the general public, the impacts of pesticides on aquatic ecosystems are less obvious, though measurable by ecologists. The cost of removing pesticides from raw drinking water supplies in the USA and Europe, however, is readily understandable and runs into hundreds of millions of dollars.

So what is diffuse pollution? It is the truly non-point source contamination (such as seepage of nitrate from agricultural land into underlying groundwaters), together with the myriad, individually minor, point sources such as forestry channels, field drains from farmland and urban surface water outfalls, that collectively deliver significant contamination to the aquatic environment. Mobilization and delivery of pollutants is often dependent on weather conditions, and may be influenced by soil type and surface cover. Simplistically, it is concerned with the individually minor, but collectively significant sources. Diffuse pollution is therefore associated with many dispersed sources, which is not to say there are no aggregations of sources of pollution within a catchment, and hierarchies of risks can often be constructed.

These characteristics make diffuse pollution problems very difficult to quantify adequately and even more difficult to control. This book aims to introduce the key aspects of the diffuse pollution problem as it affects the aquatic environment and to introduce the approaches developed in various parts of the world to control diffuse pollution.

1.2 WATER QUALITY AND POLLUTION

Water quality reflects the composition of water as affected by nature and man's cultural activities, expressed in terms of both measurable quantities and narrative statements. In many parts of the world, (e.g. the United States and European Union) the descriptive water quality parameters are related to intended water use. For each intended use and water quality benefit, there may be different parameters best expressing water quality. Both single-compound (for example, ammonia, nitrate, dissolved oxygen, phenols, etc.) and multiple-compound parameters (oil and grease, whole effluent toxicity, coliforms, etc.) are used.

Previously, water quality was expressed and perceived in terms of numerical values of various chemical and physical parameters. Today the quality status of receiving water bodies is understood in a more comprehensive manner and expressed as *integrity*. Consequently, the statutory definition of pollution included in the USA Water Quality Act Sec. 502–19 (1987) is:

The term 'pollution' means man-made or man-induced alteration of chemical, physical, biological, and radiological integrity of water.

By the same logic, the EU Water Framework Directive (Described in Chave, 2001) refers to:

... conditions for surface waters, which are defined when:

1 there are no, or only very minor, anthropogenic alterations to the value of the physicochemical (and hydromorphological) quality elements for the surface water body type from those normally associated with that type under undisturbed conditions, and
2 the value of the biological quality elements for the surface water body reflect those normally associated with that type under undisturbed conditions, and show no, or only very minor, evidence of distortion.

Today, in addition to chemical parameters, water quality assessment is more and more using and relying on biotic integrity indices that generally express the impact of pollution on composition, numbers and well-being of aquatic organisms.

As the definition of pollution given above indicates, pollution is caused by man and is reflected by downgrading the ecological integrity of the water body. Water quality comprises contributions of chemicals and other contaminants, including microorganisms, dissolved gasses, solids, and sediments. These constituents can be considered as pollution if they are in excessive concentrations that would impede beneficial uses of the water body and impair its integrity. Other human impacts not involving discharges of chemical pollutants may also cause pollution. Such human actions include hydraulic modifications of the streams, cutting down riparian vegetation lining the streams, or hydrologic modifications in watersheds that make stream channels unstable and eroding.

1.2.1 Background and natural water quality

Water draining natural habitats such as pristine forest is unpolluted; however, it contains chemicals, microorganisms and sediments. The origin of these chemicals is the contact of rainwater with the vegetation (tree canopy throughfall), soils, decaying vegetation, animal and insect droppings and others. These water quality constituents make *background* or *natural water quality*. In most cases, natural water quality represents the most pure state of surface water. There are cases, however, where natural water quality is not as good and can diminish the beneficial uses of the receiving water. For example, streams draining natural wetlands in temperate regions often have very low dissolved oxygen concentrations may affect fish populations. Anaerobic decomposition and evolution of methane, from highly organic wetland sediment which consumes oxygen, is common in stagnant waters. Other examples of natural water contamination, but not pollution by man, include high carbon dioxide content of some groundwater which is injurious to building materials, and elutriation of humic organics from decaying aquatic vegetation which impairs the suitability of water for potable supply.

The terms background and natural water quality are often considered as the same thing. Recently, however, *background water quality,* refers to water quality contributions from uncontrollable distant and regional/global sources. For example, small PCB (polychlorinated bi-phenyl) concentrations have been measured in glaciers of Greenland and Antarctica, in spite of the fact that PCBs are exclusively man-made pollutants.

Knowledge of background/natural water quality is important in diffuse pollution abatement. The same meteorological processes – rain, surface erosion, elutriation of chemicals – that form the natural chemical and biological composition of surface waters also generate pollution. The difference in some cases is the intensity at which the key water quality constituents are elutriated from the land surface of soils into the receiving waters. Natural water quality does provide a *reference* on the most desirable water quality in the region.

One of the most obvious examples of possible degradation of water quality by natural causes is the metallic content of water and sediment. Indeed, natural soils and minerals sometimes do contain potentially toxic metals, and the sediments in

Table 1.1 Chemical concentrations in Satila River (Georgia, USA) (Source Beck, Reuter and Perdue, 1974)

Summer water temperature (°C)	*Dissolved oxygen (mg/l)*	*pH*	*Aluminium (mg/l)*	*Iron (mg/l)*	*Organic matter (mg/l)*
21–27	3.3–6.0	3.8–4.6	0.19–0.6	0.83–1.6	39 –72

the aquatic system have a composition that is derived from erosion of these formations. It is quite possible that, due to the very high but natural and uncontrollable sediment concentrations in rivers that carry high natural sediment loads (e.g., Yellow River in China, some rivers in arid southwestern United States), total metal concentrations may be very high and may exceed the established criteria for metals.

Table 1.1 shows an example of the natural water quality in a pristine coastal river located in Georgia, USA. The stream is draining lowland forest and wetlands. It can be seen that some water quality constituents (e.g., dissolved oxygen, aluminium) could cause a water quality problem and could violate standing water quality standards. The cause however, is natural and, therefore, does not represent pollution and the authorities should consider adjustment of the water quality standards to reflect this reality. A *use attainability analysis* is an official document in the US that allows a variance in water quality standards to reflect natural and other non-removable water quality impairments (Novotny *at al.,* 1997). Background (natural) water quality is different in different regions. *Ecoregions* are geographical units that exhibit similar ecological characteristics, including natural water quality (Omernik and Gallant, 1990). Ecoregions do not often coincide with *watersheds* or *river basins,* which are geographical units within a watershed divide. A watershed can be a part of several ecoregions.

Environmental quality standards are concentrations of pollutants that should not be exceeded in aquatic systems if water quality characteristics are to be preserved. Environmental quality standards (EQS values) are useful parameters to classify water quality, and form a basis for assessments of assimilation capacity (Novotny and Olem, 1994).

1.3 POINT SOURCE AND NON-POINT SOURCE POLLUTION

The term diffuse pollution has evolved from an earlier recognition of two categories of pollution sources: point source and non-point source. During the last century, most pollution prevention efforts focused on clean up and control of discharges from effluents from cities and industries. These discharges are called *point sources* because they usually come from identifiable and recognizable pipes, sewer outfalls, underwater diffusers, and discharge channels. If untreated, the point sources create unsightly scum, sludge, anoxia, fish kills, and generally poor water quality of the receiving waters. The receiving waters then become unsuitable for contact recreation and most other beneficial uses. The common characteristic of point sources is that they enter the receiving water bodies at some identifiable single or multiple point location carrying pollutants. A common characteristic of major point sources is that in most countries these sources are regulated; their control is mandated and a permit is required for the waste discharges from these sources.

During the last thirty years of the twentieth century, efforts to clean point sources intensified and billions were spent on the installation of treatment plants which have significantly reduced pollution loads and improved water quality.

Researchers began to recognize the significance of *non-point source* pollution in the late 1960s and early 1970s as water-quality models and mass-balance calculations revealed significant sources of pollutants other than major point sources. Non-point sources were manifested in the field when improvements in wastewater treatment at point sources failed to produce the anticipated water quality improvements in streams and rivers. With continuing water quality improvements and control of individually significant point sources, non-point sources became increasingly recognized in Europe and North America as an important aspect of environmental water quality. The control of non-point sources is usually more complex and difficult than for point sources in that non-point sources often involve complex transport and transformation through several media (e.g., atmospheric deposition, soil application, and chemical transformation in soils involving the air, soil, and water media). Moreover, one cannot, in a practical sense, directly regulate non-point source emissions, but only activities in the watershed that may cause emissions.

There has been a growing awareness of the importance and severity of non-point source pollutants in European and American legislation. New environmental directives for water quality by the European Community impose strict limitations for a variety of non-point source pollutants (for example nitrogen). In the US non-point sources were recognized in the 1987 reauthorization of the federal Clean Water Act. Section 319 of the act requires the US Environmental Protection Agency and individual states to assemble information that characterizes non-point source pollutant impacts on receiving water quality, to report to the US Congress on the nature of non-point pollution, and to present a plan to address non-point sources.

The importance of non-point sources on water quality has been confirmed by studies in Europe and North America. Nitrogen is a pollutant that exemplifies non-point source pollution. A recent report on the Danube River basin found that non-point sources contributed 60 per cent of the nitrogen (and 44 per cent of the phosphorus) load to the entire river basin (CEC, 1994). The problem has not been solved; the European Environment Agency (1999) noted that the concentrations of nitrate in EU rivers have seen little change since 1980, leading to eutrophication in coastal areas. Also, nitrate contamination of aquifers remains an issue. In the decades since then environmentally significant reductions in loadings are still hard to find.

Examples of non-point sources of pollution in the USA are (Novotny, 2003):

- return flow from irrigated agriculture (specifically excluded from the point source definition by the US Congress),
- other agricultural and silvicultural run-off and infiltration from sources other than confined concentrated animal operations,
- unconfined pastures of animals and run-off from range land,
- urban run-off from unsewered urban areas,
- run-off from small and/or scattered (less than two hectares) construction sites,
- septic tank surfacing in areas of failing septic tank systems and leaching of septic tank effluents into groundwater,
- wet and dry atmospheric deposition over a water surface (including acid rainfall),
- flow from abandoned mines (surface and underground), including inactive roads, tailings and spoil piles,
- activities on land that generate wastes and contaminants such as:
    - deforestation and logging;
    - wetland drainage and conversion;
    - stream channelization, building of levees, dams, causeways and flow diversion facilities on navigable waters;
    - construction and development of land;
    - interurban transportation;
    - military training, manoeuvres and exercises; and
    - mass outdoor recreation.

Statutory considerations aside, the term non-point source can be misleading however, since whether or not a polluting input is a point or non-point source is a matter of scale. A field drain is a point source into a length of ditch, and the sum of discharges from an entire city can be modelled as a single point source in a river system such as the Mississippi or the Nile. Furthermore, the process characteristics of non-point source pollution – dependence on weather and land uses for mobilization and impacts, for example – are also characteristics of many minor point sources, such as storm drains (surface run-off), field drains and seasonal ditches. Therefore, the term diffuse pollution has come into existence to include minor point sources as well as non-point sources. The two terms are often used interchangeably, which may cause confusion.

Run-off from storm sewers (typically exemplified by urban developments of all kinds), concentrated animal feeding operations, and construction sites have the characteristics of both non-point and point source pollution (see below). The pollution from these sites is intermittent, occurs mostly during meteorological events, and the pollution originates from land use activities, which are characteristics of non-point sources, yet the discharge is usually through an identifiable outlet or overflow point. In the USA such sources are legally point sources, although in other parts of the world liabilities for the various pollution sources on, for example a stormwater sewer catchment, would not be legally controlled as for a major, single municipal sewage effluent or industrial process discharge. In respect of the nature of the pollutant sources, their mobilization and transport, and also issues of controllability, it is useful to have an expanded definition of *diffuse sources*, that may include both point and non-point sources.

1.4 DIFFUSE POLLUTION

Diffuse pollution comprises non-point source contamination, such as sheet run-off from fields or seepage from soil into groundwater, as well as pollution arising from a multiplicity of dispersed, often individually minor, point sources such as surface water drains in urban areas, field drains, and ferruginous springs associated with abandoned mine workings.

A practical definition of diffuse sources and pollution has been proposed in the United Kingdom as follows (D'Arcy *et al.*, 2000a):

> Pollution arising from land-use activities (urban and rural) that are dispersed across a catchment, or subcatchment, and do not arise as a process effluent, municipal sewage effluent, or farm effluent discharge.

Diffuse sources can be characterized as follows (Novotny and Olem, 1994; Novotny, 2003):

- Diffuse discharges enter the receiving surface waters in a diffuse manner at intermittent intervals that are related mostly to the occurrence of meteorological events.
- Waste generation (pollution) arises over an extensive area of land and is in transit overland before it reaches surface waters or infiltrates into shallow aquifers.
- Diffuse sources are difficult or impossible to be monitored at the point of origin.
- Unlike traditional point sources where treatment is the most effective method of pollution control, abatement of diffuse load is focused on land and run-off management practices.
- Compliance monitoring is carried out on land rather than in water.

- Water quality impacts are assessed on a catchment scale.
- Waste emissions and discharges cannot be measured in terms of effluent limitations.
- The extent of diffuse waste emissions (pollution) is related to certain uncontrollable climatic events, as well as geographic and geologic conditions and may differ greatly from place to place and from year to year.
- The most important waste constituents from diffuse sources subject to management and control are suspended solids, nutrients, faecal pathogens and toxic compounds.

1.4.1 Pollutants of concern

Traditional point source pollutants include suspended solids and their organic (volatile) content, biochemical oxygen demand, pathogenic microorganisms, and nutrients (nitrogen and phosphorus). The biodegradable organics reduce dissolved oxygen levels in the receiving water bodies and cause other nuisance problems such as accumulation of organic sludge, sediment oxygen demand and promotion of nuisance algal growths.

Pollutants of concern, originating from diffuse sources, are presented in Table 1.2. Toxic chemicals, both inorganic (metals and salts) and organic (polyaromatic hydrocarbons – PAHs and solvents) are the most serious pollutants in the run-off from urban areas and highways while sediment, nutrients and pesticide loads are most troublesome in agricultural run-off and subsurface flows. Heavy salt usage and receiving water loads during winter from municipalities and roads are serious water quality problems in snowbelt areas of North America and Europe. In addition to salt (sodium and calcium chloride), winter salt-laden snowmelt also contains complex cyanide (an anticaking additive to salt) that may breakdown to toxic hydrogen cyanide and several toxic compounds (Novotny *et al.*, 1999).

1.4.2 Diffuse pollution loads and impacts

Figures available for Scotland indicate that diffuse pollution that comes from agriculture, forestry, urban drainage and acid rain accounts for over 2,000km of polluted watercourses. If mine drainage (often manifested as dispersed springs and seepages) is included, the figure is 2,530km. By comparison, only 93km, or 2.1 per cent of Scotland's polluted rivers are due to pollution caused by industrial effluents (SEPA, 1999). Diffuse pollution also affects standing waters and reservoirs, as well as groundwaters. Diffuse pollution from agriculture in Scotland, for example, adversely affects 83 per cent of the country's polluted lochs. Urban drainage (contaminated surface run-off) accounted for 504km of polluted rivers and coastal waters in Scotland. For the UK as a whole diffuse pollution has been identified as a major unresolved pollution issue, with adverse impacts on water supply and wastewater treatment utilities, nature conservation interests, fisheries, and recreation (D'Arcy *et al.,* 2000a). According to Cunningham (1998) non-point (approximating to diffuse) sources were the principal contributors of pollutants to 76 per cent of the US lakes and reservoirs that failed to meet water quality standards. Non-point sources similarly impaired 65 per cent of the US streams failing to meet standards and 45 per cent of the estuaries (USEPA semi-annual reports).

1.5 LAND USE AND TRANSITION

Diffuse pollution by definition is primarily caused by use of land by man and transformation of land use. Land-use transformation and land-use categories are depicted on Figure 1.1. There are four types of pre-development lands: prairies (steppes), wetlands, forests and arid lands, including deserts and semideserts. The conversion activities include: slash burning of forests; cultivation of prairies and forests cleared by burning; drainage and filling of wetlands; irrigation; construction

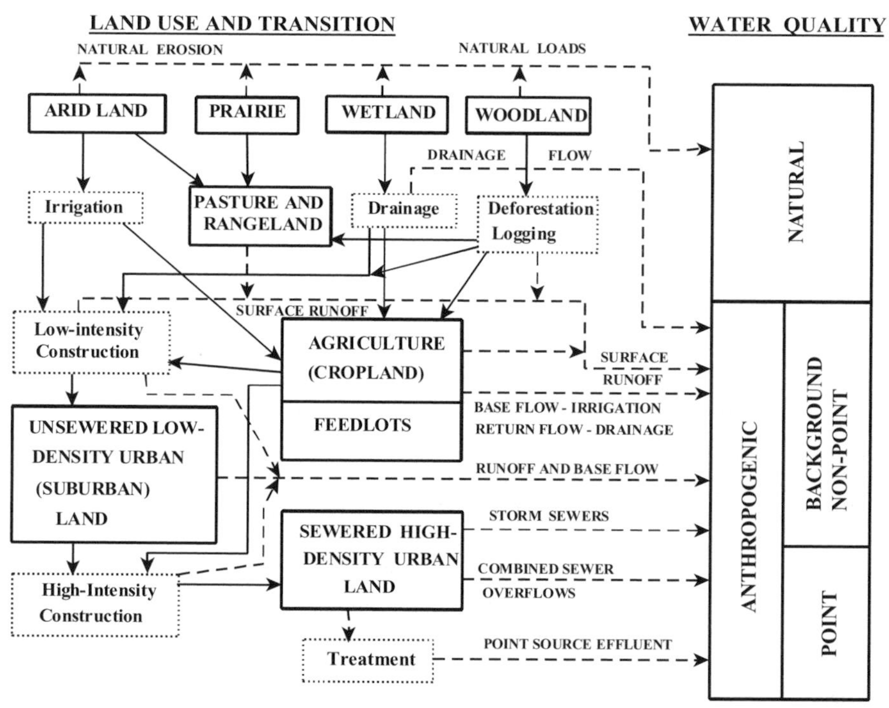

Figure 1.1 Land use and land use transformation, types of diffuse pollution and effects on water quality (from Novotny and Olem, 1994 and Novotny, 2003)

Table 1.2 Diffuse pollution concerns in the United Kingdom (from D'Arcy *et al.*, 2000a)

Pollutant	*Example source*	*Environmental problem*
Oil and other hydrocarbons including:	Car maintenance. Disposal of waste oils. Spills from storage and handling. Traffic emissions and road run-off. Industrial emissions.	Toxicity. Contamination of urban stream sediments. Groundwater contamination. Nuisance (surface waters), taste (potable supplies).
Solvents	Cleaning on industrial yards, illegal connections of industrial effluents to surface waters.	Toxicity. Contamination of potable supplies (rivers and groundwater).
Pesticides	Municipal application to control roadside weeds. Agriculture. Private lawn care.	Toxicity. Contamination of potable supplies.
Suspended solids	Run-off from arable land. Upland erosion. Accumulation of solids on impervious urban surfaces. Construction.	Destruction of gravel riffles. Embeddedness. Sedimentation in natural pools and ponds. Carrier of nutrients and toxic compounds.
Biodegradable organic wastes	Agricultural wastes (feedlots, silage liquor, surplus crops). Sewage sludge. Land application of effluents.	Oxygen demand. Nutrient enrichment.
Faecal pathogenic microorganisms	Septic tank system failures. Animal faeces in towns and cities. Illegal cross-connections of separate sewer systems. Applications of organic wastes to land.	Health risks. Noncompliance with recreational standards (beach closings).
Nitrogen	Agricultural fertilizers. Traffic emissions. Atmospheric deposition.	Eutrophication (especially of coastal waters). Contamination of potable supplies. Acidification.

(continued)

Table 1.2 continued

Pollutant	*Example source*	*Environmental problem*
Phosphorus	Soil erosion. Agricultural fertilizers. Contamination of urban run-off (detergents, organic materials).	Eutrophication of freshwaters: • ecological degradation • increased potable water treatment cost • nuisance algal growths.
Toxic metals	Urban run-off. Land application of industrial and sewage sludge.	Toxicity.
Acidity of atmospheric precipitation	Car emissions. Power plant emissions.	Toxicity. Aesthetic nuisance.
Sodium and cyanide	De-icing chemicals for winter road traffic safety.	Toxicity.

activities; surface mine excavation, etc. Each conversion process produces pollution and pollutant loads during the conversion process that are typically several orders of magnitude higher than the background loads from the original native land. Deforestation and large scale construction are the most polluting land-use transformations.

Land conversion and removal of mangrove wetlands is a problem with international consequences in tropical and subtropical countries. Rapid development of shrimp production – driven mostly by demand in developed countries – has led to the destruction of mangroves for shrimp ponds. These ponds have a limited life-span and after being abandoned, leave the land unusable and polluted (Tonmanee and Kanchanakool, 1998).

Another major problem is deforestation and land conversion to intensive uses such as agriculture and urban development. Deforestation is caused by unsustainable logging (clear-cutting) for commercial lumber and wood. In developing countries, deforestation is driven by population growth, limited soil fertility, land tenure inequities (Sanchez, 1992) and by foreign demand for cheap wood and wood products.

Eighty per cent of tropical deforestation is caused by non-traditional cultivation. At the subsistence level traditionally practised in forests for millennia by indigenous peoples, shifting cultivation may be sustainable. Small-scale farmers clear and burn a few hectares of land a year, mainly to grow food. When the fertility of the reclaimed soil is exhausted the farmer clears another area and the previously cultivated land is allowed to stay fallow for 10 to 15 years and left to reforest. Problems arise when populations increase, often associated with migration following road construction for timber extraction.

The fields created by shifting cultivation in tropical forests typically have low yields; problems with fertilizer and pesticide application are non-existent. But more land is required when population increases occur. For example, under shifting cultivation, upland rice farms in Peru and the Amazonian regions that were created by deforestation, will yield about 1 tonne per hectare of crops, while for conventional farming rice yields would be about 11 tonnes per hectare.

The rapid degradation of soil organic matter content due to high temperatures and the breakdown of the soil structure due to tillage, leads to a rapid decrease in the hydrologic functions and crop productivity. For example, in Indonesia, Abujamin and Abujamin (1985) reported that three to four years of annual food crops under typical tillage practices lead to decreases in land productivity. The rapid decrease in soil productivity raises an issue of the so-called desertification of tropical lands. It is true that formerly lush, productive tropical lands have been converted into wastelands in a few years (Melching and Avery, 1990). Loss of productivity leads to demand for more land, which increases deforestation, siltation and diffuse pollution.

Tropical forests on Borneo are disappearing at a rate nearly twice that of the Amazon (Duff-Brown, 1999). The reason is mostly commercial logging for export of cheap wood. The island is shared by Malaysia and Indonesia. Logging on the island began in the 1970s and Malaysia quickly became the world's top exporter of tropical hardwood for scaffolding, chopsticks and furniture. Logging provides a livelihood for about 100,000 families but, at the same time, displaces the aboriginal population. The logging is carried out in mountainous tropical forests and the consequences include increased erosion and siltation of receiving waters. In 1991 the International Tropical Timber Organization warned that Borneo's Sarawak forest would be denuded by logging in the short period of thirteen years, if the numerous timber concessions did not cut production drastically. It recommended halving exports and halting logging on steep slopes to prevent erosion.

Governments have partially responded. Exports were reduced and the government developed plans and restrictions on logging. The logging system allows for the felling of eight to ten trees for every hectare of the forest, replanting, and then allowing the forest to regenerate for 15 years. About 12 per cent of the forest has been set aside as protected areas for national parks and wildlife sanctuaries. In spite of these actions, claims have been made that erosion and river silt have already destroyed the ecological balance of streams and forests, including habitats for orang-utans.

Shifting cultivation with subsequent reforestation may be possible in humid tropical countries. Traditional shifting cultivation, with low population densities and soil erosion which is not excessive, allows forests to grow back (Sanchez,1992). However, in countries with relatively high population densities, lower precipitation, and farming on erosive higher slopes lands, e.g., Haiti, Ethiopia and several other African countries, deforested land has lost top soil and the regeneration of forests has not occurred. From history it is known that land deforested in the Middle Ages by Venetians in the mountainous Dalmatian region of Croatia has not returned to forest.

Besides soil loss, conversion of native lands to agriculture changes the soil chemistry, which may lead to a significant loss of chemicals. For example, changes of prairies into arable lands and the drainage of wetlands triggered nitrification of large amounts of organic nitrogen stored in the native soils, and released large quantities of nitrate into groundwater and subsequently into the base surface flow (Kreitler and Jones, 1975). Deforestation, conversion of deforested lands to agriculture, and erosion of agricultural lands are the main pollution problem in developing tropical and subtropical countries.

The scale of land-use change in developing countries pales into relative insignificance when compared with the developed world. In the UK, for example, woodland cover declined from about 80 per cent of pre-developed land area to only 4 per cent by 1940. The state of Illinois 150 years ago contained large wetlands that covered a significant proportion of the state. Most of the wetland area was drained and converted to monocultural (corn) agriculture. Until very recently, efforts to halt or reverse the deforestation trend in the UK during the twentieth century typically involved extensive monoculture plantations of exotic tree species. Such land use changes were often associated with significant loss of nutrients resulting from the ploughing up of upland and other hitherto permanent pasture land, as well as siltation of watercourses and leaching of trace elements such as aluminium salts from naturally poorly buffered soils. Forests represent only a part of the story of land-use change, of course, as agriculture has developed from its initial origins in the Near East and then the Mediterranean countries, to the Agricultural Revolution which occurred in the UK in the seventeenth century, and then the development of the agro-chemical industry in the industrialized world from 1940 especially. The UK as a case study does not end there, unfortunately: intensification has continued. Figures 1.2 to 1.5 show land-use changes from 1930–1995. The associated changes in chemical fertilizer and pesticide use have been well documented (European Environment Agency, 1999). There has also been an increase in the population of livestock in the UK. The sheep population has seen the most dramatic recent changes, increasing from 22 million in the 1940s to 44 million in the UK by 1993, with predictable consequences for grazing land (Sansom, 1999b).

The UK also well illustrates the consequences of urban land-use changes in terms of diffuse pollution. As the home of the Industrial Revolution, the UK became

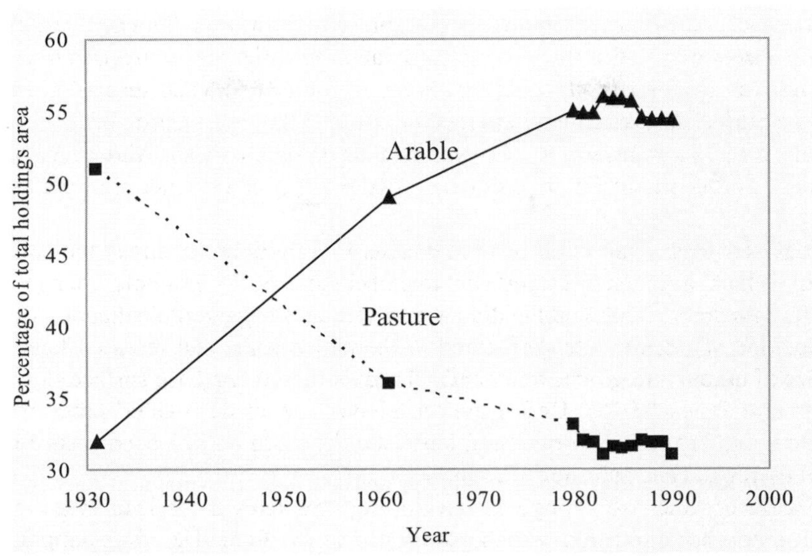

Figure 1.2 The change in arable and pasture balance in England (RSPB quoted in Environment, Agency 1999)

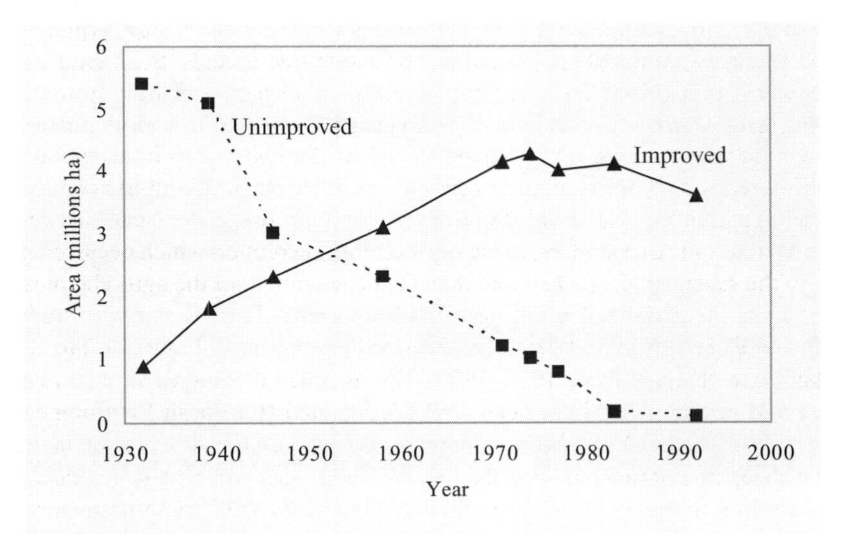

Figure 1.3 The increase in improved pasture and the decrease in unimproved pasture in England (RSPB quoted in Environment Agency, 1999)

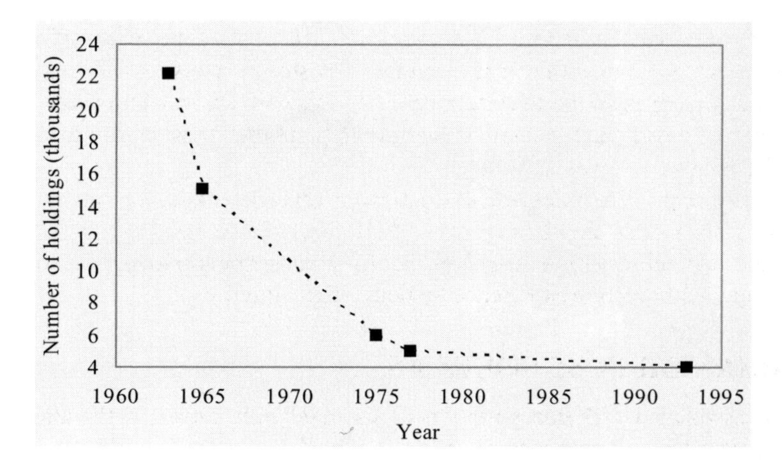

Figure 1.4 The loss of general mixed farms in England and Wales (Environment Agency, 1999)

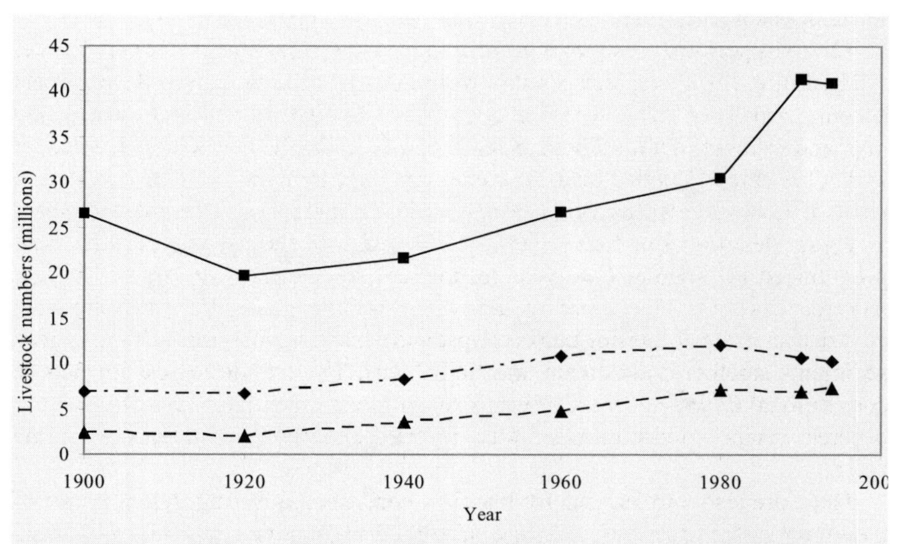

Figure 1.5 The loss of general mixed farms in England and Wales (Environment Agency, 1999)

the most urbanized country in the world, with several cities with populations of a million or more in the nineteenth century. Further industrial expansion occurred during the middle decades of the twentieth century, plus new urban phenomena: ribbon development along trunk roads and the development of extensive suburban land around the fringes of towns and cities.

The USA similarly illustrates the dramatic impacts of land-use changes; particularly the history of the dust bowl years of the 1930s, which led directly to the formation of the soil conservation initiatives that have subsequently been able to offer best management practice experience to many other countries.

1.6 URBAN DIFFUSE SOURCES

Urban impacts associated with transportation are especially important: in the UK road traffic increased 63 per cent between 1980 and 1996.

Diffuse pollution from urban sources has been extensively covered in the literature (Novotny and Olem, 1994; Marsalek and Torno, 1993; ASCE-WEF, 1998; Marsalek *et al.,* 2001; Novotny, 2003). The most comprehensive research on the magnitude of urban run-off pollution was conducted in the US under the auspices of the USEPA which sponsored the National Urban Run-off Project (USEPA, 1983).

The most profound changes that affect the loading of pollutants from urban sites are hydrological. Land surfaces that were mostly pervious before development become partially or fully impervious as a result of urban development. Consequently rates and volumes of run-off both increase significantly after development. Schueler *et al.* (1992) reported that peak flow rates can increase by a factor of 2 to 10 as a result of urbanization. Roesner (1999a) presented a comparison of pre-development and post development flows and documented that the differences are most pronounced for smaller (two-year recurrence) storms than for large (100-year recurrence) storms. The higher frequency of peak flows causes the urban stream to cut a deeper channel, causing bank collapse and destroying marginal habitat. Eroded sediments smother downstream aquatic habitats. These hydrological impacts of conventional drainage are the reason why so many urban rivers are protected in concrete channels, and streams are often culverted, resulting in no amenity or wildlife value.

There are also implications for low flow conditions. The impervious nature of the urban surface prevents soils and subsoils from receiving rainfall infiltration. Incident rainfall is conveyed swiftly into the receiving watercourse by conventional urban drainage systems. During periods of prolonged dry weather there can be little residual seepage of water from the urban catchment to sustain low flows. Once again, poor habitats for wildlife and fisheries result, and the near dry watercourse attracts litter from the human population.

Urban sources of diffuse pollution include (Novotny and Olem, 1994; Novotny, 1995; Marsalek and Torno, 1993; ASCE-WEF, 1998; Novotny, 2003):

1. Atmospheric deposition that can be wet and dry
2. Traffic emissions and vehicle and road wear out
3. Urban erosion of idle lands and construction sites
4. Pet and wild animal/birds faecal matter
5. Fertilizer and pesticide applications on urban lawns
6. Winter ice- and snow-removal practices
7. Leaking septic tanks in unsewered urban areas
8. Illicit discharges into storm sewer systems such as spent oil from vehicles, detergents from car washing and other illegal dumping
9. In developing countries, practices of discharging grey wastewater and other wastes into surface drainage systems
10. Contaminated inflow/infiltration into storm sewers mostly from cross-connection with sanitary sewers or illicit discharges of sewage into storm drainage
11. Erosion of drainage channels
12. Build-up of solids in sewers, especially in combined sewers during dry periods, resulting in anaerobic conditions for storm discharges.

The amount and character of the pollution and the load depends on the type of surface on which the pollutants accumulate during a dry period and on the drainage of the urban area. Urban drainage systems contain remnants of, or designed natural drainage, and conduits that can be either open channels or underground combined or separate sewers. In the separate sewer systems, household sewage and industrial wastewater is carried by *sanitary (or foul) sewers* and urban surface run-off is conveyed by *storm (or surface water) sewers.* Natural drainage for conveyance of surface run-off is now implemented in newer lower density developments in parts of the USA and increasingly in Europe too (see Chapters 3 and 4). This type of drainage, if it is properly designed (best management practice), is typically able to handle once in five- to ten-year run-off events and provide run-off treatment at the same time.

Combined sewers are far more common in the much older cities of the rest of the world, and in older sections of US cities, (mostly in the central and eastern states). Combined sewers carry a mixture of sewage/wastewater (dry weather flow) and storm/rainfall run-off. Their capacity to handle flows is limited typically to four to eight times the dry weather flow. The excess flow is relieved from the system by *combined sewer overflows (CSOs)*. The quantity and frequency of the CSOs are now regulated and limited. According to some European and emerging US CSO control ordinances, typically, 90 per cent of the overflow volume is to be captured, stored and treated before it can be released into the receiving water bodies. Without such controls, the CSO discharges are a serious local pollution problem.

Surface drainage channels (masonry lined or unlined) are common in some developing countries and their function is a cross between storm and combined sewers. These channels carry all stormwater and grey household wastewater from surrounding housing.

Loading rates are defined as the wash-off quantity of a specific pollutant over a period of time per unit area of total watershed. Event mean concentrations (EMC) are generally used where time-related run-off quantities and qualities are not available or statistically well-defined. An EMC is a flow-weighted average concentration of a pollutant in a storm run-off (flow) or snowmelt event. Run-off water volumes are then used to convert the concentrations to mass. The loading rate per unit area of a watershed can be estimated on a yearly basis using:

$$L = 0.01\, R \times C$$

where L = annual loading, kg/ha/yr
R = annual run-off, mm/yr
C = EMC, mg/l (= g/m^3)
0.01= conversion constant, (kg/l)/(ha/mg/mm)

For urban stormwater run-off the statistical properties of the EMCs were extensively studied in the US by the Nationwide Urban Run-off Project (NURP) (USEPA, 1983). The study has found that values of the EMCs were log-normally distributed for the major urban land uses (residential, commercial, and typical industrial). Their statistical characteristics (log mean and log standard deviation) were not statistically distinguishable between the typical urban land uses (except open vacant or parkland). In addition the EMCs were not correlated to typical independent variables (parameters) such as storm characteristics, geographical location, soil type and others. This implies that at least in concept, the EMCs loading models would be applicable to conditions outside of the US where land uses are more mixed.

Table 1.3 shows the nationwide average statistical characteristics of the EMCs as obtained by the NURP project. Loading rates have been calculated for various applications using the data from the USEPA's (1983) Nationwide Urban Run-off Program (NURP) (Cole *et al.,* 1984), the Federal Highway Administration (Gupta, 1981), and other site-specific studies (e.g., in Orlando, Florida by Wanielista *et al.,* 1981). The NURP study found that the run-off coefficient (volume of run-off divided by the volume of rainfall) which affects the run-off volumes is strongly correlated, as one would expect, to the percent imperviousness of the area. Furthermore, rainfall volumes are geographically variable, consequently the loads of pollutants will differ between the land uses and geographical locations.

Table 1.3 Overall water quality characteristics of urban run-off from NURP studies (after USEPA, 1983)

Constituent	*Typical coefficient of variation*	*Site median EMC*	
		For median urban site	*For 90 percentile urban site*
Total suspended solids (mg/l)	1–2	100	300
BOD_5 (mg/l)	0.5–1	9	15
COD (mg/l)	0.5–1	65	140
Total phosphorus (mg/l)	0.5–1	0.33	0.70
Total kjeldahl nitrogen (mg/Ll	0.5–1	1.5	0.21
NO_{2+3} – N	0.5–1	0.68	1.75
Total Cu (μg/l)	0.5–1	34	93
Total Pb (μg/l)	0.5–1	144	350
Total Zn (μg/)	0.5–1	160	500

From the data measured and reviewed in the NURP studies, it was evident that metals formed one group of chemicals that were consistently above the detection limit. Many priority pollutants were below the detection limits or were infrequently detected. As an example, 71 priority pollutants were detected in the NURP urban samples. All 13 priority metals were detected. Copper, lead, and zinc were found in at least 95 per cent of these samples. The organic toxic pollutants were detected at a much lower frequency. Of the 57 organics detected, 46 were present in only 1–9 per cent of the samples. There were 106 sampled organic compounds. Some of the aromatic hydrocarbons related to gasoline or oil products were detected at very low concentrations and frequencies. Preliminary data from the municipal stormwater permitting programs indicate that metals' concentrations were lower than reported in NURP (CH2M-Hill, 1990; CDM, 1993), particularly for lead which can be attributed to the elimination of the use of leaded gasoline in the US.

Ellis *et al.* have published some equivalent data for the UK, and identified oil as a major contaminant of urban sources (in D'Arcy *et al.,* 2000a).

1.7 AGRICULTURAL DIFFUSE SOURCES

Until the 1950s, most farming, even in Europe and the USA, was carried out on smaller family farms that used organic fertilizers and, essentially, their waste production was easily assimilated by soils and receiving water bodies. Since the 1950s there has been a worldwide shift to larger monocultural, intensively operated farm units. Farm yields have increased dramatically; however, to sustain increasing yields and productivity, farms are using large quantities of chemical fertilizers and pesticides. At the same time, deforestation has occurred on a large scale since the 1950s and the deforested land has been converted to agricultural (mostly in developing countries) and urban (mostly in developed countries) land uses.

Diffuse pollution from rural lands can be categorized as follows:

- increased erosion and soil loss
- chemical pollution
- irrigation
- livestock

1.7.1 Increased erosion and soil loss

According to Melching and Avery (1990) 80 per cent of serious erosion problems are now occurring in developing countries. Planting on steep slopes, or other highly erodable land is common in developing countries and, as a result of the small size of fields, terracing is a traditional and very effective form of erosion control that has been practised for centuries, with apparently minimal adverse environmental consequences. On the other hand, land that has been recently deforested, either for commercial logging or converted subsequently to agriculture, erodes at an accelerated rate. Typically, the sediment loads of streams draining recently (in the last thirty years) deforested watersheds have increased by several orders of magnitude. As pointed out in the preceding section, most of the increase of erosion in developing countries is caused by deforestation.

With the exception of arid lands, soil loss by erosion from fields is several orders of magnitude larger than the background load. Soil loss is caused by the conventional ploughing and soil cultivation practices that keep the soil bare during cultivation and planting. Some row crops such as corn do not provide enough vegetative soil cover. Soil erosion is the major cause of diffuse pollution and sediment is the most visible pollutant. The environmental impacts of excessive erosion and sedimentation caused by agriculture can be listed as follows (Clark *et al.*, 1985):

1. Effects of excessive sediment loading on receiving waters include deterioration or destruction of aquatic habitats. Excessive deposition of sediments in slow-moving reaches and impoundments blankets the bottom fauna, 'paves' the bottom of the streams, and destroys fish spawning areas. Excessive sediment loads can also directly harm fish and other aquatic wildlife.
2. Excessive sedimentation causes a rapid loss of storage capacity in reservoirs and accumulation of bottom deposits that inhibit normal biological life. In many parts of the world, reservoirs built on sediment-laden streams were filled in a matter of years, sometimes before the full function of the reservoir is achieved.

3. Nutrients carried by sediment can stimulate algal growth and, consequently, accelerate the process of eutrophication. Phosphates and, to a lesser degree, ammonium from fertilizer application and pollution discharges are adsorbed by soils and suspended sediments.
4. Fine fractions of the sediment are primary carriers of other pollutants such as organic toxic compounds and metals. For example, persistent organochlorine compounds, such as aldrin and dieldrin pesticides, have low solubilities in water but are readily adsorbed by suspended sediment.
5. Turbidity from sediment reduces in-stream photosynthesis, which may lead to reduced food supply and habitat. In littoral zones of reservoirs and shallow sections of rivers, increased turbidity causes a shift from rooted aquatic plants to planktonic algae that usually has adverse water quality and habitat consequences.

In many countries, many fields are found in flood plains; large-scale soil losses are associated with periodic major floods. Silt generated by excessive erosion tends to clog drainage canals and further encourages seasonal flood plain farming. High concentrations of salts in the agricultural seasonal floods encourage growth of weeds after the floods have passed (Agrawal, 1999).

1.7.2 Chemical pollution by fertilizers and pesticides

The most rapid adverse changes in environmental degradation occurred in developed countries and countries of the former Soviet bloc after the 1950s and accelerated in the 1970s and 1980s. Streams and lakes in agricultural areas, which before the 1960s were reasonably clean, are now suffering from excessive algal growths and other symptoms of eutrophication caused by discharges of nutrients from fields and animal husbandry. Groundwater that was previously safe for drinking is now unsuitable for human consumption due to high nitrate content and contamination by organic chemicals, some of them carcinogenic. The most severe water quality changes caused by excessive applications of agricultural chemicals have occurred in Central Europe, the Netherlands, the United Kingdom, and some parts of North America.

In the agricultural sector, pesticide use has increased both in industrialized and developing countries. In India pesticide use has increased nearly fifty-fold between 1958 and 1975, yet Indian consumption in 1973–1974 was reported to be averaging about 0.33 g/ha, compared to 1.48 kg/ha in the USA and 1.87 kg/ha in Europe (Avcievala, 1991; Ongley, 1996). The largest use of these chemicals is in western Europe. A further increase was noticed between 1975 and 1990.In 1993, the Netherlands used about 20 kg/ha of organic chemicals, Belgium 12 kg/ha, France 6 kg/ha and Germany 4 kg/ha (UNIDO, 1984). Figure 1.6 shows historical trends in the use of industrial fertilizers in several countries.

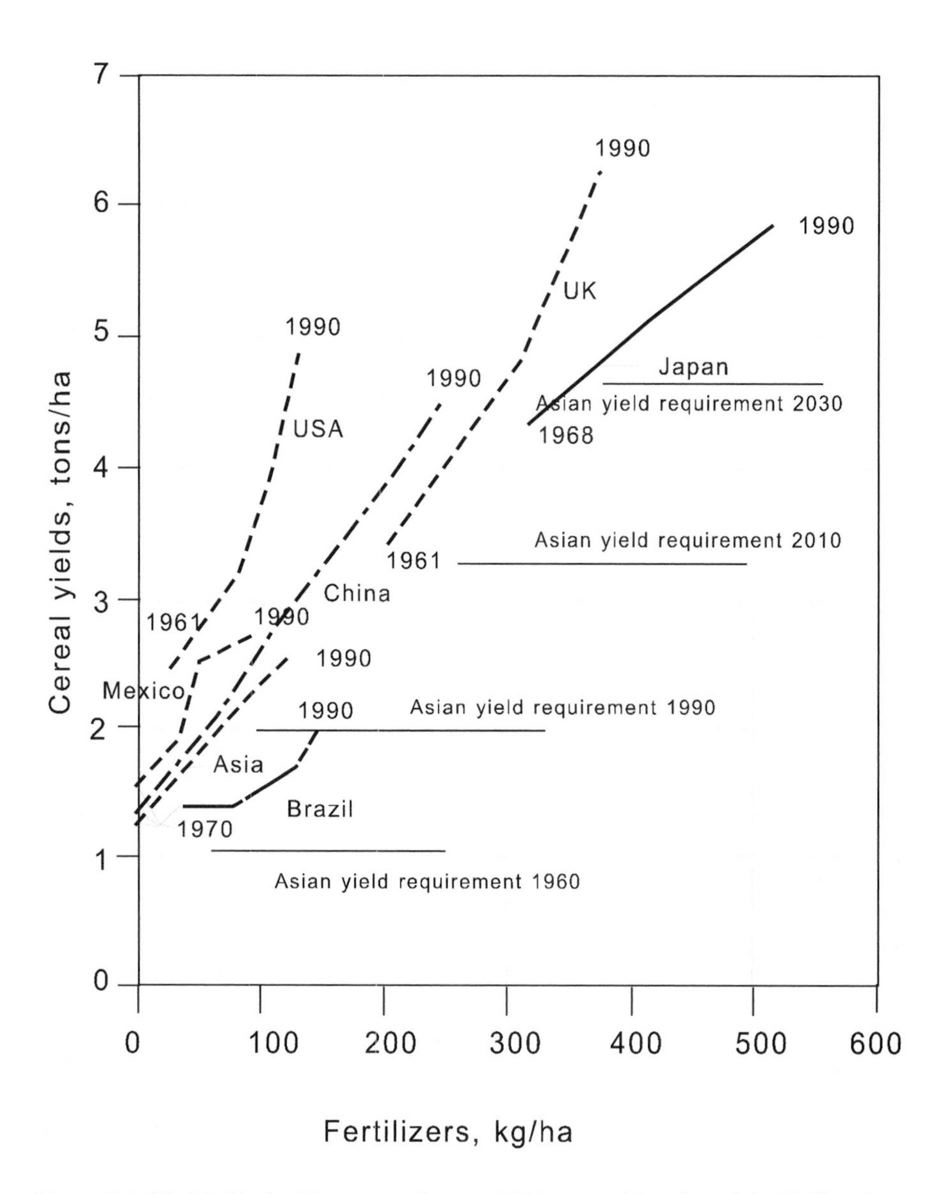

Figure 1.6. Worldwide fertilizer use and crop yield increase (data from Joly, 1993 and Ongley, 1996).

Sources of nutrients from agriculture can be categorized as livestock sources and emissions from fields. Livestock wastes accounted on average for 30 per cent of the total phosphorus load in European inland waters, and the rest of agriculture accounted for additional 17 per cent (ECE, 1992). Waste from concentrated livestock operations are considered as point source pollution.

Tonmanee and Kanchanakool (1998) listed Thailand as the 20th largest consumer of pesticides worldwide and fifth among developing countries, after China, Brazil, Algeria and Egypt. Most pesticide use is to control insects in fruit orchards. The authors also pointed out one serious consequence of pesticides use. In most cases, citrus orchards have irrigation canals that carry pesticide laden irrigation return flows. Consequently fish in these canals are highly contaminated and pose a health hazard to those who consume the fish.

Similarly to other more advanced countries, the use of pesticides in agriculture in China has also dramatically increased. The average rate of pesticide application in China today is about 3 kg/ha of active ingredients (Li, 1999). This value is higher than application rates reported for India and is comparable to rates reported from US and Canada.

1.7.3 Irrigation and irrigation return flow

To satisfy the ever-increasing demand for food, intensive irrigated agriculture is increasing in developing countries. As pointed out in the preceding section, irrigation return flow, a consequence of irrigation, is a pervasive type of diffuse pollution. Irrigation return flow is the hydrological excess of applied irrigation water over the evapotranspiration requirement. If excess water is not applied, salts build up in the soil and to the detriment of agriculture (as has occurred in several parts of the world).The irrigation return flow moves across the surface and through the topsoil to a receiving water body. It contains salts, ions and nutrients leached from the soil and from excess chemicals applied as fertilizer.

1.7.4 Pollution from animal husbandry

Pollution from animal husbandry can be divided into that from pastures and that from concentrated animal operations (feedlots). Cattle being reared for meat are scattered in pastures, whereas milking on dairy farms means that the cows are gathered in a concentrated area several times a day and yards, though small (typically less than one hectare), produce very high pollution loads. Studies of loads from yards located in Wisconsin, Michigan and Ontario by Moore *et al.* (1979) provide information on nutrient loads: phosphorus production by one dairy cow is 18 kg/year.; and of that amount, a significant portion may reach the receiving waterbody.

Table 1.4 Typical concentrations of pollutants in farmyard run-off and pasture

	Pollutant concentration, mg/l				
	BOD_5	*COD*	*Total N*	*Total P*	*Source*
Feedlot run-off	1,000–11,000	30,000–40,000	920–2,100	290–380	Loehr (1972)
Grazed pasture	n/a	n/a	4.5	7	Robins (1985)

This load depends on the proximity of the farm to the watercourse and on the pollution attenuation during overland flow.

Run-off from animal husbandry, if it is not controlled, can severely affect receiving waterbodies. The most obvious impact is the depletion of dissolved oxygen caused by a high BOD of the run-off. The BOD concentrations of farmyard run-off exceed that of sewage by more than twice as shown in Table 1.4. The run-off from farmyards also carries pathogenic microorganisms, including protozoa *Cryptosporidium.* In developing countries animals are kept by farmers in small numbers typically allowed to roam freely on pastures, in the countryside, near and in the water bodies and in settlements.

Overgrazing in arid regions of developing countries, such as in southwest Asia (and also in the southwest United States) causes environmental degradation of watersheds and plant cover loss that can also be classified as diffuse pollution.

Extreme example of upland erosion, Scotland. (photo: Alan Frost)

Megacities are a severe example of gross pollution from diffuse sources, as a result of a lack of adequate drainage infrastructure. Johannesburg, South Africa. (Photo: Manda Hinsch)

The application of pesticides presents risks of water pollution.

Run-off from yards at livestock units, industrial premises, roads, and car parks is an important pollution source.

Individually minor sources can be collectively significant (example illustrations: car washing, outdoor storage of 'empty' drums).

Clear felling timber can result in the mobilization of nutrients and suspended solids, with severe ecological impacts.

2

Best management practices

2.1 INTRODUCTION

For diffuse sources best management practices (BMPs) are the regulatory and practical means of controlling water quality. They are the building blocks that enable water resource planners to manage water pollution risks and ensure acceptable standards in a waterbody. Thus, conceptually, they are the pollution control equivalent of end-of-pipe treatment plants (with associated environmental agency individual quantitative permits) for major point source effluent discharges. The need for such a distinction is further considered in the Section 2.2.

The following definition of best management practices was published in Novotny and Olem (1994, p. 18):

> Best Management Practices (BMPs) are methods, measures, or practices selected by an agency to meet its non-point (diffuse) source controls needs. BMPs include, but are not limited to, structural and non-structural controls and operations and maintenance procedures. BMP can be applied before, during, and after pollution-producing activities to reduce or eliminate the introduction of pollutants from diffuse sources into receiving waters.

The USA recognized the problem of diffuse pollution many years ago and established provisions in an amendment to the Clean Water Act in 1987, leading to national programmes of action to address the issue (Weitman, 1996). The concept of best management practices was developed as a practical answer to diffuse pollution problems from all sources and sectors. The best management practice concept, as developed in the USA, therefore provided a list of defined measures for forestry, agriculture, urban development, and also for marinas and related problems such as river engineering.

The best management practice concept has several key elements:

- There is a need for guidance that offers practical prevention options.
- The options need to be defined and explicit – best practice rather than ill-defined individual interpretations of what is required.
- In order that the options can be described as best practice, they should be based on research and experience.

For each target sector (e.g. forestry, agriculture, built environment, etc) there are two classes of best management practices: (a) procedures and (b) structures. For most situations, a suite of best management practices is available. Combinations of good procedures with provision of one or more structures are not unusual. Example lists of BMPs are given in Section 2.3 below and an introduction to the more frequently encountered or broadly applicable BMP structures is provided in Sesction 2.4. A glossary of terms is included as Appendix 2.

The central importance of the best management practices approach to US efforts to control diffuse pollution is evident from the size and scope of the US EPA (1993) publication *Guidance Specifying Management Measures for Sources of Non-point Pollution in Coastal Waters:* five sector-based chapters, supported by chapters on wetlands, riparian areas and vegetative treatment systems, plus monitoring and tracking techniques; totalling 846 pages and describing scores of specific BMPs for all sectors. A useful introduction to the guidance is given in Frederick and Dressing (1993). The US EPA was required to develop 'Management Measures Guidance' by section 6217(g) of the Coastal Zone Authorization Act Reauthorization Amendments (CZARA) of 1990. That legislation required the EPA to specify management measures that were:

> Economically achievable measures for the control of the addition of pollutants from existing and new categories and classes of non-point sources of pollution, which reflect the greatest degree of pollutant reduction achievable through the application of best available non-point pollution control practices, technologies, processes, siting criteria, operating methods or other alternatives.

The management measures were developed by determining the most effective methods that were economically achievable for the control of non-point source (diffuse) pollution. For each management measure, a range of methods and practices are described that can be used to achieve the level of control expected of the measure, the applicability of the measure to sources and locations, the pollutants that can be controlled by the measure, factors to take into account in applying the measure to any given site, the effectiveness of the measure, and costs.

Guidance thus developed in the USA has been taken up and modified for local conditions in many other countries subsequently. The requirements for evidence of cost effectiveness, of local applicability and provision of a suite of practices, are what make the BMPs approach different from many national or sector-based codes of practice and similar advisory materials.

In the UK a well-developed best practice control strategy is in place for forestry: guidance in the industry's *Forests and Water Guidelines* publication (Forestry Authority, 2002) is encouraged via grant aid for forest developers (Nisbet, 1994, and Nisbet *et al.,* 1998 reviews the effectiveness of the measures therein).

2.1.1 Limitations of the best management practices approach

The problem of diffuse pollution requires more than a shopping list of BMP options for effective and sustainable control. The challenges are considered in Novotny and Olem (1994) and Novotny (2003), and papers in successive diffuse pollution conference proceedings (Olem, 1993; Straskraba ,1996, Novotny and D'Arcy, 1999; Burkart and Heath, 2002). The range of controls has been briefly reviewed in D'Arcy *et al*., 1998, and includes education, economic instruments, and proscription of the use of persistent pollutants. The latter option would greatly reduce maintenance costs for urban BMPs, for example, if quantities of toxic metals present in urban and highway runoff could be reduced at source. Similarly, for the rural environment, the banning of chlorinated hydrocarbons for use as agricultural pesticides and their replacement by products that are degradable in the soil has been a major improvement. Better guidance, development of adequate funding schemes that could apply nationally, and education about the problems and solutions are all needed. Additional means to control diffuse pollution must, therefore, include root causes as well as solutions of the symptoms of diffuse pollution. This wider consideration of the means to control diffuse pollution is considered in more detail in Chapters 7 and 8. The rest of this chapter concerns the options available to water quality planners and sector managers to work to minimize diffuse pollution by the use of BMPs.

2.2 OPTIONS TO CONTROL POLLUTION – WHY DIFFUSE POLLUTION REQUIRES A DIFFERENT APPROACH

Point source discharges are typically controlled by issuing a discharge consent or authorization which permits the discharge, subject to various conditions designed to ensure that no pollution occurs as a consequence. Discharges such as sewage effluents, or process effluent from particular industries, can usually be characterized by the presence of a small number of polluting constituents, such as BOD and ammonia in sewage, for example, or iron and suspended solids in minewater drainage. The volume of the discharge is typically well known and can be regulated to control pollutant loads. Thus by adopting maximum pollutant concentrations for the aquatic environment, (environmental quality standards, EQS) consistent with appropriate uses of the watercourse, discharge limits can be determined using available dilution.

2.2.1 Is an EQS approach appropriate for diffuse pollution control?

An EQS approach, for an individual discharge, requires the following:

I. One or two key pollutants which can be modelled and which will accurately reflect the properties of the discharge (clearly some industrial discharges are more complex than this and present additional challenges)
II. A robust environmental quality standard for each of the key pollutants.
III. A defined point of impact and simple relationship of discharge flow with available dilution (for example greater receiving flow provides greater dilution for a soluble pollutant such as ammonia).

For many diffuse pollution sources, there is no single point of discharge, or so many that setting individual controls at each point would be impractical (field drains from farmland, for example). There is also little point in setting specific numerical standards for an outlet (e.g. a field drain, urban surface water outfall, or a drainage channel from a forestry plantation) if there is no responsible organization managing (a) the process that produces the pollutants and (b) the consequent tailor-made process that reduces the pollutants to acceptable levels (as determined by EQS values for the receiving watercourse).

For a forestry channel, for example, the mobilization of pollutants will be a function of slope (topography) planting and harvesting (broadly predictable risks associated with the known technologies), soil type and rainfall. Sediment is usually the principal pollutant, but since sediment is also a natural constituent of rivers and streams there are difficulties in developing a satisfactory EQS value to control discharges.

What about the urban environment? The practicalities are similar in principle to those that have led to widespread use of best practice strategies for forestry and agriculture:

- Diffuse, multiple sources; many surface water outfalls with diffuse sources of pollution across their catchments.
- Pollutant loads are a function of storm events, discontinuous and difficult to model with confidence (see Figure 2.1)
- Pollutant loads from individual sources are not under the day-to-day control of an individual organization
- EQS values are often inadequate for protection of a catchment from urban developments (especially in relation to sediment and oil)
- Even where a reliable EQS value does exist for an urban pollutant –trace metals and some pesticides, for example – how does the regulator know that controlling that pollutant alone will effectively prevent pollution from the urban run-off?

One of the most important urban run-off pollutants is oil, for which no EQS exists (Ellis and Chatfield, 2001). Others are suspended solids and trace metals.

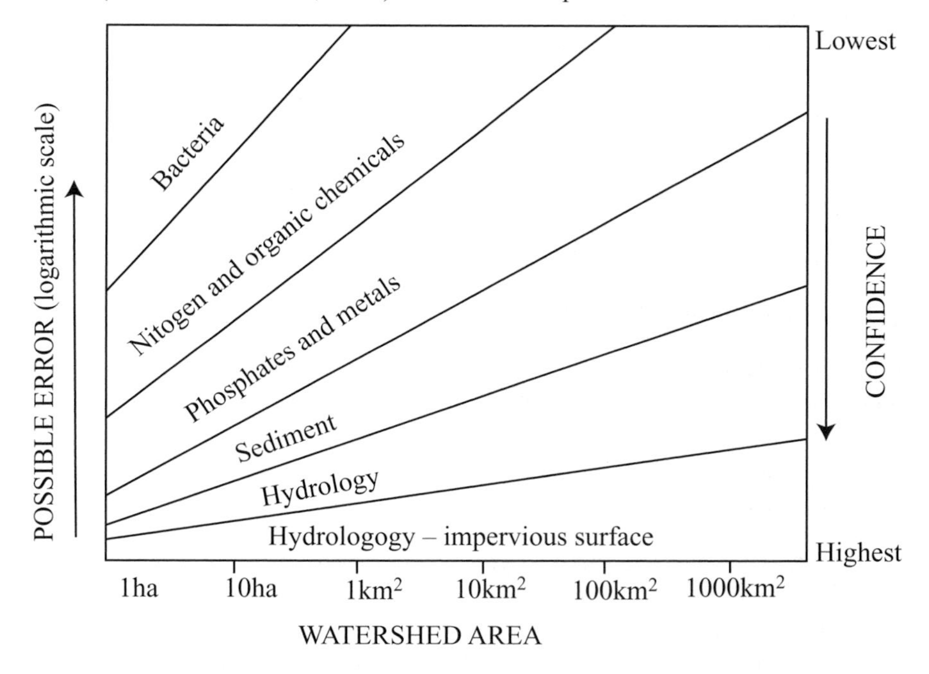

Figure 2.1 Relative accuracy and reliability of hydrologic models of diffuse pollution. Accuracy and reliability decreases with the increased complexity and size of the modelled system. (Novotny and Olem, 1994)

The latter are typically associated with sediment and likely to accumulate in river and local sediments, to be possibly released when summer redox conditions change, or be mobilized during major storm events. Simple compliance of water column samples for most of the year with published EQS standards will not mean that there is no pollution, since sediment contamination by adsorbed oil, trace metals, or pesticides will adversely affect the aquatic invertebrate fauna.

The absolute bureaucratic impracticality of individually controlling countless diffuse, minor point sources (whether field drains, forestry drainage channels, or individual surface water drains from urban developments) also precludes a conventional EQS and discharge consent approach. Where the EQS approach cannot be made to work, some alternative means of control has to be found. A code of management measures involving recognized, demonstrably effective best management practices is a practical option. For diffuse pollutants for which national or local EQS values have been set, it is then possible, by reference to anticipated performance of management measures and monitoring representative examples, to attempt to model and control diffuse pollution on a catchment level (rather than by individual discharge). A best practice approach can therefore be advocated as the only practical way of controlling urban run-off, just as it is the most practical option to control many aspects of forestry and agricultural activities.

Best practice can be sought by, for example (a) direct regulation, using descriptive consents or authorizations for major drainage schemes, or general binding rules on a sector by sector basis; (b) planning requirements set by local authorities in consultation with others; (c) trade association specifications; (d) local stakeholder buy-in to community-based river catchment plans. Regulatory and other options to control diffuse pollution are considered in Chapter 7.

2.3 HOW DO BMPS WORK?

BMPs are technology-based systems that ideally are source-control techniques, rather than end-of-pipe treatment options. They are often grouped into two broad categories:

a) procedures
b) structures.

Procedures include everything from asking dog owners to collect their pets' faeces for safe disposal, to detailed instructions for loading and unloading oil and chemicals at a factory, or the application of pesticides to a crop. Structures are facilities built to engineering standards to contain pollution risks or to actually intercept or retain pollutants. Best practice typically involves both these categories of measures for the effective control of diffuse pollution. It is perhaps more useful to consider how measures can prevent pollution and illustrate how combinations of practices can be effective.

The non-point (diffuse) source pollution control process recognizes two opportunities for intervention:

1 *Source control* is the first opportunity in any control effort. Controls vary for different types of problems, as exemplified below.
   - Reducing or eliminating the introduction of pollutants to a land area, for example reduced nutrient and pesticide application.
   - Preventing pollutants from leaving site during land-disturbing activities. Examples include using conservation tillage, planning forest road construction to minimize erosion, siting marinas adjacent to deep waters to eliminate or minimize the need for dredging, and managing grazing to protect against overgrazing and resultant increased soil erosion.
   - Preventing interaction between precipitation and introduced pollutants. Examples include installing gutters and diversions to keep rainwater away from contaminated surfaces, diverting rainfall run-off from areas of land disturbance at construction sites, or timber felling areas for example, and timing chemical applications or logging activities based on weather forecasts or seasonal weather patterns.
   - Protecting riparian habitat and other sensitive areas. Examples include protection and preservation of riparian zones, shorelines, wetlands, and highly erodable slopes.
   - Protecting natural hydrology. Examples include the maintenance of pervious surfaces in developing areas, riparian zone protection, and water management.

2. *Delivery reduction* is often necessary in addition to source controls. It entails the interception of mobilized pollutants, by capturing the runoff or infiltrate and passing flows through some sort of best practice treatment system. Examples (described below) include grass swales, infiltration basins, detention areas, retention ponds and stormwater wetlands, grass filter strips, and buffer zones.

As indicated above, a hierarchy of measures can be recognized; BMPs work by:

1 minimizing the introduction of pollutants to the environment;
2 by avoiding mobilization of pollutants; and
3 by reducing transfer of pollutants from the land to the aquatic environment.

The original guidance from the USEPA (1993) in the USA was a sector-based listing of practices, with a lot of duplication between sectors (justifiable on the basis that readers may only look at their own sector). A great deal of elaboration and conceptualization quickly followed, as various sectors and the academic world took up the BMPs approach. These are briefly described below.

2.3.1 Urban BMPs

In an urban context, the following hierarchy of practices is now usually advocated (e.g. D'Arcy and Roesner, 1999, and CIRIA, 2000):

- housekeeping (a range of business-specific measures to control pollution risks at source);
- stormwater source control (infiltrating or collecting rainwater run-off as close to the point of contact with urban surfaces as possible);
- site controls (measures to allow rainwater to infiltrate or be detained to allow removal of pollutants in suspension);
- regional controls (end-of-system features such as retention ponds and wetlands – see below) .

Housekeeping measures include contingency plans for accidents such as can occur for example when heating oil is delivered to a school or hospital, or chemicals to industrial premises. They include instructions and guidance for staff or households to minimize pollution in the course of their routine lives and businesses, for example by not washing vehicles on impervious roads or yards that drain directly to surface water sewers. Housekeeping measures may be engineered structures that contain risks of pollution – for example placing oil storage tanks within sealed bunds. It is possible on commercial premises to bring many diffuse pollution risks together and hold them within one installation designed to cope with potential accidents. This is shown in Figure 2.2, based on surveys of incidents that identified a range of causes of pollution from minor spills and drips to leaks, larger spills and failures with transfer pipes or pumps, to major incidents involving overfilling tanks or burst tanks (Usman *et al.*, 1998). Table 2.1 lists housekeeping BMPs for householders in residential areas.

The remaining categories of urban BMPs in the list above are all structural BMPs – stormwater drainage features, such as swales, ponds, permeable surfaces etc. A description of such features is given in Section 2.4, and design and selection guidance for the built environment is set out in Chapter 3.

2.3.2 Rural BMPs

The USEPA (1993) manual gives details of hundreds of practices; the aim here is therefore just to give a sample of rural options. More examples are provided in Chapters 5 and 6, where the appropriateness of particular practices and combinations thereof is considered in relation to water-quality issues..

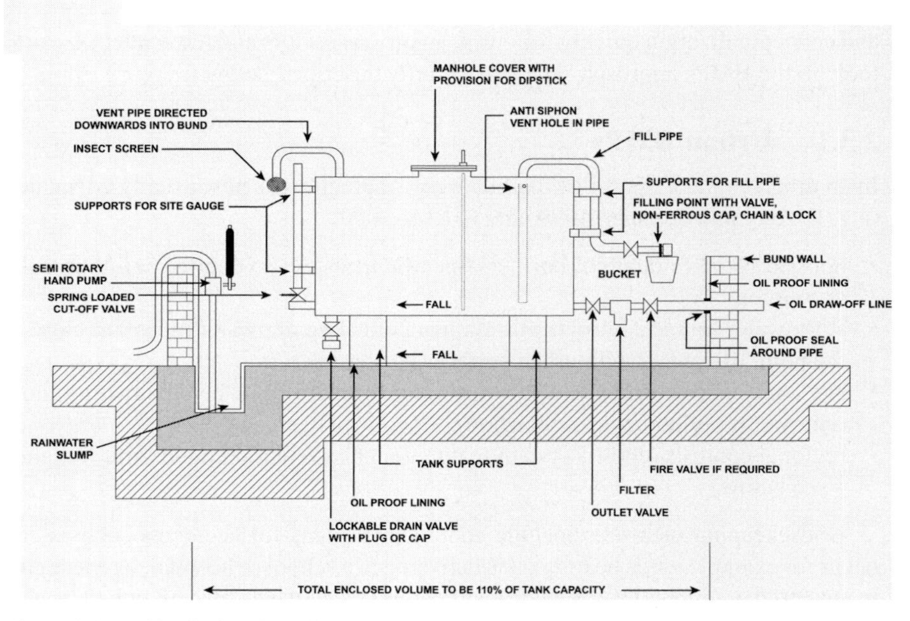

Figure 2.2 An ideally bunded oil storage tank.

Table 2.1 Examples of urban good housekeeping practices (for individual householders)

Best management practice	*Potential pollutant control*
Collection of pet faeces for safe disposal	Faecal pathogens, nutrients, BOD
Litter collection from street gutters	COD in gulleys and drains, large material that may cause blockages, plastic, metals
Recycling waste oil	Oil and solvents
Car washing on grass or unmade ground or at a commercial facility drained to adequate treatment or recirculation facility	Oil, detergents, grit
Recycling or safe disposal of paint thinners and paint residues, disinfectants.	Solvents, oils and chemicals
Minimizing use of garden pesticides and disposing excess in hazardous chemical collection facilities	Pesticides

BMPs : a four point focus

Planning tools	In-field measures	River margins	Built environment
Nutrient budgets	e.g. Conservation tillage, grassing run-off carrying depressions in fields	Buffer zones	Swales and retention ponds or wetlands
Manure application plans	Crop residue mulches	Fencing off livestock	Roof cover to exclude rainfall from dirty yard areas
Pesticide procedures	Field drainage maintenance	River restoration	Biobeds for pesticides
Contingency plans for accidents	Irrigation scheduling		
	Locating access tracks for livestock, and feedlots away from watercourses		
	Grazing management		
	Stocking densities		

Figure 2.3 Example BMPs for agricultural businesses to protect the aquatic environment.

In a rural context it is useful to consider how BMPs can address pollution risks across the business. Again, a different classification of constituent practices is helpful in assessing risks and pollution control options on a farm. Figure 2.3 presents this schematically.

A similar clustering of BMP options is set out for forestry in Figure 2.4. More information about the example measures in these figures is given in Chapter 5, which considers in more detail aspects of BMP design and selection for rural environments, and the effectiveness and limitations of such approaches is considered in Chapter 6. Descriptions of some of the more frequently encountered structural BMPs – rural and urban – are given in the following section.

For any sector – for example farming, forestry or urban – it is obvious that a suite of measures, selected for appropriate circumstances, will be needed. Each successive measure will further reduce risks of pollution being a significant problem. In the urban context this is *the treatment train* concept (D'Arcy and Roesner, 1999, and Ellis, 1982 – and see Chapters 3 and 4). In a farming context, farmers would be expected to look at appropriate BMPs across the range of options in Figure 2.3, not

Planning Considerations	Drainage	Minimising soil disturbance
Indicative forestry strategy for area including EPA water quality sensitivities (nutrients, siltation, acidifying precipitation) critical load assessment	Blind ditches for furrows	Mound planting
Likely period to harvest	Silt traps (ponds)	Contour planting
• Nutrient needs?	Roadside swales	Use of straw bales and brush at harvest
• Pest control strategy	Interception ditches	Buffer zones of natural vegetation
Slopes and planting	Leaving natural drainage pathways intact and with natural vegetation cover and not overshaded	Timber extraction techniques
Harvesting methods and timing	Location and construction of access roads	

Figure 2.4 Forestry BMPs for protecting water quality – summary classification

just rely on in-field measures, for example, nor to simply hope that a buffer strip will adequately control pollution risks without good pesticide practices, nutrient budgets and in-field measures. Risks need to be considered at each of the four key components in the BMPs suite in Figures 2.3 and 2.4 and similarly for other sectors. Combinations of measures across the range of risk activities are needed.

2.4 EXAMPLES OF STRUCTURAL BMPS AND LANDSCAPE FEATURES

Structural BMPs are facilities built to recognized design standards, for which published performance data is available. They vary in application and generally function in one or more of the following ways:

- Stabilize ground surfaces to reduce risks of pollutants being mobilized, or chemicals that are not pollutants when in the soil becoming pollutants when washed into a stream (e.g. nutrients and soil particles).
- Permit infiltration of rainwater into the ground as close to source as possible, and reduce risks of run-off scouring material from farm, highway or urban surfaces.

- Move intensive agricultural, or urban/industrial activity away from water margins by establishment of riparian buffer zones.
- Intercept pesticide spray drift, and/or stabilize soils from wind erosion, or provide shelter for livestock away from watercourses (these measures would not always be near a watercourse) by establishment of landscape features in fields.
- Detain stormwater run-off for short periods (24–48 hours) to permit the deposition of suspended matter and associated other pollutants.
- Filter out suspended matter, thereby intercepting and retaining pollutants that have been mobilized by rainfall or other factors, prior to the pollutants reaching a watercourse.
- Retain run-off in retention ponds or constructed wetlands long enough (2–3 weeks, CDM, 1993) for biodegradation of organic pollutants (e.g. faecal pathogens, detergents, oil and other substances) and uptake of nutrients by algae and macrophytes.

Examples of the ground stabilization measures are the establishment of grass swales in arable farmland along the courses of hollows and channels that become ephemeral watercourses in wet weather, the provision of seed-impregnated biodegradable matting to protect newly landscaped slopes in urban and some industrial developments (to reduce risk of soil erosion), and livestock paths that use absorbent materials such as wood chips or bark, or fine stone chips where underlying soil will allow drainage.

Infiltration techniques can include grass swales alongside roads, tracks and yard areas (in the built environment especially – including farms) if on permeable soils; gravel filter drains alongside roads; individual soakaways for buildings; infiltration basins; and permeable pavements for car parks and pedestrianized areas in urban developments. Figure 2.5 shows a cross section diagram of a permeable pavement of a type widely used in the UK in recent years.

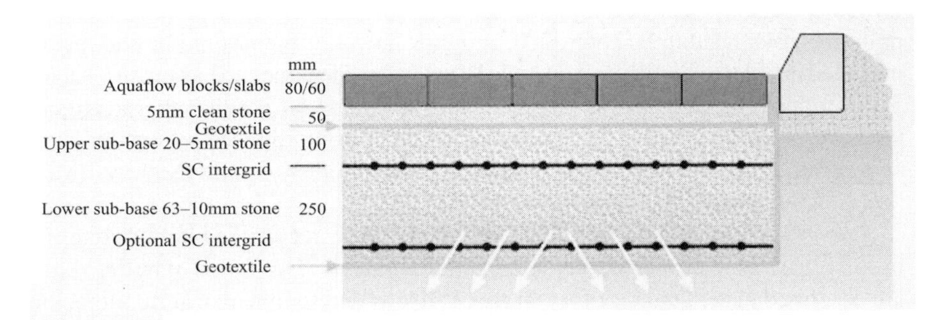

Figure 2.5 Diagram of a permeable surface car park to show drainage (K.Hart, Formpave).

For agriculture, heavily engineered specific structures are more unusual. Often a simple change in practice can facilitate infiltration or in other ways minimize mobilization of potential pollutants. For example, contour ploughing on arable farmland reduces the risk of rapid flows of run-off downslope that can entrain soil and lead to erosion and water pollution.

There are a number of structures or features that are widely used to help minimize pollution from various sectors, and their presence in the landscape has amenity and wildlife potential that needs to be taken into account in planning these BMPs. In many instances landscape measures are strongly advocated, not just to play a role in the control of diffuse pollution, for example in many urban design guides (e.g. Campbell and Ogden, 1999; Ferguson, 1998), but also in rural locations too, for example in the American publication of the Committee on Long-Range Soil and Water Conservation, *Soil and Water Quality: An Agenda for Agriculture* (1993). They include buffer zones, wetlands and ponds; all valued for their potential to be aesthetically pleasing landscape features and for their potential wildlife interest, as well as two other widespread practices: swales and contour planting. The rest of this section introduces these key features that are useful in dealing with run-off in many contexts. Design details are given in the Chapters 3 and 5.

2.4.1 Buffer zones

The buffer zones idea has received a lot of research attention, perhaps as a result of an excessive expectation of their pollution-control capabilities, or perhaps driven by the reluctance of causal sectors to allocate land. They include major landscape features such as broad belts of woodland, as well as more modest features such as hedgerows and dry-stone walls that trap soil erosion off fields, and grass filter strips. The idea is simple – to allocate land for a zone between the potentially polluting activities (e.g. farming, forestry or urban development) and the at-risk watercourse, in order to minimize risks of pollutant transfer from the land to the water. The difficulty begins when considering what width of buffer zone and what types of pollution risks need to be controlled, and thus how the zone should be set up and managed in order to be effective. Detailed design considerations are set out in Chapter 5 and, for now, an idea of aims and characteristics follows. Like any other treatment measure, the effectiveness of a buffer zone can be compromised by overloading, and buffer zones are best used in conjunction with other, complementary BMPs (see Chapters 5 and 6).

The 1993 US guidance, reproduced in part below, recognized buffer zones as useful practices in farmland, forests and urban contexts. Two forms of buffer zones are described in the guidance for control of erosion and sediment in agriculture, the first being a means of holding back run-off from arable land and the second a means of filtering it as it passes from the fields towards the watercourse.

Field border: A field border is a strip of perennial vegetation established at the edge of a field by planting or by converting it from trees to herbaceous vegetation or shrubs.

Field borders serve as 'anchoring points' for contour rows, terraces, diversions, and contour strip cropping (see Section 2.4.4). By not ploughing the lower ends of slopes, erosion from long rows and concentrated flow in furrows may be reduced. This is also a recognized technique for forestry in the UK. On steeper slopes or the more erodable soils, once rills are formed the flows will cut through and in-field soil stabilization measures are required.

Filter strips (see Section 2.4.3) are often deployed as riparian measures to trap pollutants moving over the surface of the ground towards a watercourse, but not all buffer strips seek to have a filtering capability. A riparian hedgerow, especially if on a bank that has been slightly raised above the level of the adjacent field, may be effective at intercepting run-off, but may be even more important in presenting a barrier to direct overspraying of small headwaters watercourses, and also intercepting spray drift.

A lot of research has been undertaken concerning the effectiveness of different types of buffer zones for various purposes for which some benefit could reasonably be expected (e.g. Dillala *et al.*, 1989a, 1989b; Dillala *et al.*, 1988; Haycock *et al.*, 1997; Meals and Hopkins, 2001). There has even been research to see if buffer zones would be ineffective in situations where it is predictable that they would be ineffective, for example where the attention is focused on a soluble pollutant such as nitrate that can by-pass the buffer zone via field drains or sub-surface interflows (Puckett, 2004; Addiscott, 1997), although even there, there are circumstances where buffer zones can be effective (e.g. Spruill, 2004).

The importance of buffer zones for forestry too is clear in the USEPA (1993) guidance (Forestry Management Measure *Streamside Management Areas*). The measures are to control pollution that could otherwise arise from disturbance of soil (loss of nutrients, increase in turbidity in receiving waters, as well as maintaining shade to prevent streams becoming too warm for native game fish. The US guidance advises that streamside management areas should have a minimum width (35–50 feet), and that the width should increase according to site-specific factors: slope, class of watercourse, depth to water table, soil type, type of vegetation, and intensity of management. The guidance also includes the following additional measures applicable in the riparian buffer zone (stream management area):

Minimize disturbance that would expose the mineral soil of the SMA forest floor. Do not operate skidders or other heavy machinery in the zone.

Locate all landings, portable sawmills, and roads outside the zone.

Restrict mechanical site preparation in the zone, and encourage natural vegetation, seeding, and hand planting.

Limit pesticide and fertilizer usage in the zone; buffers for pesticide application should be established for all flowing streams.

The effectiveness of these and other forestry BMPs in the USA is reported in US Department of Agriculture (1994). Buffer zones are also a key part of the UK equivalent guidance document: *Forest and Water Guidelines* (Forestry Authority, 2002).

Finally, it should be noted that buffer zones are potentially important for amenity and biodiversity: corridors of semi-natural vegetation through intensive farmland or forest plantations, or through urban conurbations and industrial land. Detailed consideration of stream buffer zones in urban environments is given in Schueler, 1995.

2.4.2 Ponds

The published literature variously refers to a range of types of pond, often with significant inconsistencies between interpretations of terms. That lack of consistency explains why published data on the performance of 'ponds' is so variable and drawing conclusions can be difficult. The most important terms for which there are design criteria and published performance data, are detention and extended detention basins, retention ponds, and stormwater wetlands. Some authors merely refer to wetlands, and some swales can be developed into open, broader, wet marsh areas that certainly have wetland vegetation but do not have the residence time for pollutants of a stormwater wetland. Older references often simply refer to wet and dry ponds, where the latter are detention basins and the former have a permanent body of water – but, unless designed as a retention pond, a wet pond does not necessarily have easily predicted pollutant removal capabilities. More detailed parameters on design of ponds are in Chapter 3.

Schueler *et al*., (1992) define detention basins as follows:

Conventional ***Extended Detention (ED) ponds*** *temporarily detain a portion of storm water run-off for up to 24 hours after a storm, using a fixed orifice. Such extended detention allows urban pollutants to settle out. The ED ponds are normally 'dry' between storm events and do not have permanent standing water.*

Enhanced ED ponds *are designed to prevent clogging and resuspension. They provide greater flexibility in achieving target detention times. They are equipped with plunge pools near the inlet, a micro pool at the outlet, and utilize an adjustable reverse-sloped pipe as the ED control device.*

Wet Ponds are generally defined simply as

permanent water bodies that don't drain down completely, but have permanent water storage.

That's not a very specific description, however, and fails to distinguish detention from retention ponds (see Chapter 3). Design (sizing) is recommended on the basis of the US 'NURP pond guidelines'. The latter are said to also be the basis of the Metropolitan Washington Council of Government's manual, (Schueler, 1987), Federal Highway Administration (Dorman *et a*l., 1988) and State of California (CDM, 1993).

- The key design criterion is the 'volume ratio', the ratio of pool storage volume to mean storm volume
- 'Mean storm volume' is a statistical measure expressing run-off volume associated with the long-term average rainstorm quantity (USEPA, 1986)
- The USEPA produced total suspended solids removal curves for different parts of USA. Removal of other pollutants is proportional to suspended solids removal (to very variable extent).
- Generally volume ratio of approximately 2.5 is necessary to achieve 75 per cent TSS reduction, (at which total phosphorus would be only 50 per cent). An asymptotic relationship existed for increasing pollutant removal with increasing ratio values greater than 2.5.

For example, for total phosphorus, 60 per cent removal is approached only if volume ratio approaches 5 (discussion in Horner *et al.*, 1994).

Pools with volume ratio values of approximately 5 would have 2–3 weeks of 'pool storage hydraulic residence time' and consume 3–7 per cent of contributing catchment (dependent on impervious area, slopes, rainfall characteristics, etc (Walker, 1987; Hartigan, 1989; Kulzer, 1989). Such ponds are therefore usually now referred to as retention ponds to distinguish them from the shorter residence time permanent pools often provided in enhanced extended detention basins. The latter offer improved removal of suspended matter, but do not have the retention time needed for degradation of pollutants in solution or suspension, nor to allow adequate uptake of nutrients by algae and higher plants, as occurs in retention ponds. Detailed design guidance is given in Chapter 3.

Schueler *et al.* (1992) define an additional category of designed pond as follows:

*Conventional **stormwater wetlands** are shallow pools that create growing conditions for marsh plants. They're designed to maximize pollutant removal through wetland uptake, retention and settling. They are constructed systems and typically not located within delineated natural wetlands.*

Stormwater wetlands usually require a somewhat larger surface area than equivalent retention ponds, but can offer greater auxiliary benefits, for example, for wildlife.

The above notes are about ponds or detention basins that typically are engineered and have designed outlets for routine flows. On permeable soils, it is possible to design infiltration basins that hold stormwater only so long as it takes for the water to seep away into the ground.

On farms and in forestry areas, ponds are often merely detention areas with no routine outlets, allowing water to seep into the ground, or to overflow and wet the surrounding ground when precipitation exceeds evaporation, with excess run-off seeping into surrounding soil and eventually into groundwater or surface waters by indirect routes.

An introduction to design considerations for BMP ponds of all kinds is given in Chapter 3.

2.4.3 Swales and filter strips

A similar confusion of meanings is common here too. A swale is described in Schueler *et al.*, 1992 as a natural depression or wide shallow ditch used to temporarily store and route or filter run-off. As an urban run-off BMP, that publication states that grassed swales are conveyance systems in which pollutants are removed from urban stormwater by filtration through the grass and infiltration through soil. An urban swale should include a filter strip and operate as a source control measure. These essentially ephemeral watercourses are also widely used in agricultural and forestry practices, although in agriculture, their pollutant filtering capability is not a primary consideration, being secondary to the aim of preventing bed erosion in the course of storm flows over soil surfaces. The original USEPA (1993) guidance specifying management measures for NPS pollution from agricultural sources specifies as BMP number 412:

> *Grassed waterway: A natural or constructed channel that is shaped or graded to required dimensions and established in suitable vegetation for the stable conveyance of runoff.*

The guidance notes that grassed waterways may reduce the erosion in a concentrated flow area, which may result in the reduction of sediment and associated substances delivered to receiving waters. Vegetation may act as a filter in removing some of the sediment delivered to the waterway, although that is not the primary function of a grassed waterway. The guidance (USEPA, 1993) also notes that grassed waterways can also be pathways for the delivery of nutrients and pesticides to receiving watercourses unless additional practices are employed. The most effective complementary measures in such circumstances would probably be in-field measures, but where a filter strip is provided alongside the grassed waterway it offers additional benefits (further pollutant removal capability) and becomes closer to the sort of appearance intended in urban literature for a swale (eg Schueler *et al.*, 1992, Horner *et al.*, 1994, CIRIA, 2000). Filter strips (as agricultural measures) are described in (USEPA, 1993) as

> *Filter strips: grass cover designed to receive runoff as sheet flows from adjoining land surfaces.*

and

> *A strip or area of vegetation for removing sediment, organic matter, and other pollutants from runoff and wastewater.*

Forestry practices also recognize the use of swales and filter strips, the former to stabilize beds of ephemeral watercourses and filter strips for slopes receiving runoff into the swale. Even better than trying to establish new vegetated slow-flow drainage channels with some filtration capacity (swales), the UK *Forest and Water Guidelines* advocate the zoning of watercourses (including ephemeral flows) from the outset (planning the afforestation scheme) as untouched natural vegetation. As for some agricultural situations, in forestry it is often better not to provide any drainage pathway that could facilitate the collection, concentration and delivery of contaminants to a watercourse (even a swale). It is preferable to use turnouts or other means to disperse road and other runoff into the forest land. As well as reducing water pollution risks, such practices are especially desirable for substances that are not pollutants if left on the land such as soil, fertilizers and faecal material from livestock deposited on paths or around watering points.

2.4.4 Contour planting

The design of combinations of BMPs for specific example circumstances in rural situations is considered in chapter 5. For agriculture, these are just a handful of a whole range of practices advocated (primarily for soil erosion control) in the US guidance, but are included here as examples.

> *Contour farming: Contour farming is farming sloping land in such a way that preparing land, planting, and cultivating are done on the contour. This includes following established grades of terraces or diversions*

This practice reduces erosion and sediment production, therefore less sediment and associated pollutants are transported to receiving watercourses. Increased infiltration however, may increase the movement of soluble pollutants into groundwaters.

> *Contour orchard and other fruit areas: Contour orchard and other fruit areas involve planting orchards, vineyards, or small fruits so that all cultural operations are done on the contour.*

Contour orchards and fruit areas can reduce erosion, sediment yield and pesticide concentrations in run-off. The efficacy of this approach is not just supported by work in the USA, of course, but by thousands of years of history in the birthplaces of agriculture and civilization in the Near East and China, and subsequently all over the world. The US guidance for BMPs states for the above practice that the effectiveness of the measure depends on the slope of the bench and the cover. With inward sloping benches, the sediment and associated chemicals are trapped against the slope and 100 per cent trap efficiency may be achieved. Where planting benches slope outwards, losses are greater and the lower bench may be subject to erosion from outflows from higher ones. Soluble pollutants may leach into groundwaters.

Various forestry practices also allude to the idea of contour planting and related practices. For road systems in forests (USEPA, 1993):

> Forestry site preparation practices include constructing beds along the contour, avoiding connecting beds to drainage ditches or other waterways, and operating planting machines along the contour to avoid ditch formation. Similarly design *of roads* and skid trails to follow natural topography and contour reduces the amount of cut and fill required, and reduces potential for failure of the road. Contour trails high up the hillside reduce risks of crossing major streams and provide maximum intervening undisturbed land to accept runoff *and thereby* reduce soil erosion and encourage revegetation.

The above are just some examples of BMPs in use; details of many more are given in the USEPA guidance document referred to throughout this section (USEPA, 1993), and useful summaries are given in Novotny and Olem, 1994 and Novotny, 2003.

2.5 WATERSHED (CATCHMENT) MANAGEMENT

2.5.1 Integrated approaches

There is no single alternative that would accomplish all tasks needed in management of diffuse pollution causes and its symptoms because the pollution loads come from many sources and directions affect the water bodies (surface and groundwater) in multiple ways. Diffuse pollution control is multi-level hierarchical, typically involving several units of government agencies and non-government organizations

(NGOs). Diffuse pollution should be using *watershed-wide* approaches and management. A multi-level, watershed-wide approach deals with the problem comprehensively and combines the most effective BMPs in an effective strategy to reduce pollutant loads and restore the affected water bodies.

Watershed management is multi-objective and diffuse pollution control is one of the goals. For example, management of smaller and medium-size urban streams today must consider several objectives such as (Novotny *et al.*, 2000)

I. Flood control.
II. Preservation and restoration of the ecological integrity of the receiving water body affected by point and non-point discharges and changes in hydrology and hydraulics.
III. Providing contact and non-contact recreation for urban populations.
IV. Other uses such as water supply, navigation, or hydropower production.

Some of the objectives are conflicting and others are complementary to each other. For example, preservation and restoration of ecological integrity and providing habitat for aquatic life complements the recreational objective. As a matter of fact, the healthy ecology of a water body is a prerequisite for the contact recreational use. On the other hand, conventional approaches to flood control are often in conflict with ecological and recreational objectives in the context of watershed and water body management. These conflicts can be reconciled and uses must be optimal.

A programme to address and control diffuse pollution and other objectives of watershed management should be carefully planned and have the following important characteristics (Novotny and Olem, 1994; ASCE-WEF, 1998):

- *Comprehensive* – the programme should have adequate resources to address all aspects of the diffuse pollution sources and their water quality and water body impacts.
- *Clearly established goals and criteria* – the basic objective of diffuse pollution management is watershed-wide improvement in water quality and enhanced beneficial use of the receiving water bodies. However, not all watersheds and water bodies require extensive and expensive diffuse pollution watershed management. Extensive, integrated diffuse pollution management is required for receiving water bodies and watersheds that have a clear diffuse pollution problem. Such watersheds can then be *targeted* for comprehensive management and control. If the water quality is impaired mostly by point sources, other tools are available that rely on permitting, with the cost borne by the dischargers. Common sense and good watershed housekeeping should be practised in all watersheds irrespective of whether or not they are targeted.
- *Equitable* – each responsible party must be defined and specific directives must

be given to each party. Direction for each party should be commensurate with its share of the problem and equitable to its share of the solution.

- *Continuous and dynamic* – meaningful diffuse pollution controls will take a long time and the outcomes are not immediate. Therefore, diffuse pollution management programs should be implemented in a manner that addresses the dynamics of an evolving, long-term programme.
- *Integrated* – point and non-point sources, flood control and other objectives of watershed management should be addressed at the same time. This enables implementation of transferable pollution discharge permits and load allocation trading where such practices are permitted, and development of multi-objective controls and best management practices.

2.5.2 Quantitative aspects of catchment management for diffuse pollution control

The ultimate indication of effectiveness of pollution control activities in a catchment is provided by the ecological indicators – the fauna and flora that have to endure anthropogenic impacts. This approach is at the heart of many water quality classifications, including the EU Water Framework Directive (European Union, 2000) that sets out to define good ecological status as the target quality of water bodies.

That is especially appropriate for the impacts of many types of diffuse pollution (e.g. urban run-off into small streams, or nutrient loads from agriculture and forestry impacting lakes). But for some pollutants, it is necessary to quantify loads and compare with target concentrations in the receiving water body. Those target standards are the environmental quality standards (EQSs) referred to in Section 2.1. and in Chapter 1.

The objectives of water quality planning, including diffuse pollution control, are to achieve and protect the physical, chemical and biological quality of the receiving bodies of water and ensure the water resource is fit for its uses. Contamination by a specific substance does not constitute pollution where the contaminants are at a level below threshold values of significance for the substance and the type of water body. It follows that there is an allowable level of contamination for many substances and a capacity therefore in a receiving water body for a finite loading of contaminants: the *loading capacity* (LC). This concept is discussed in Novotny (2003). The LC comprises the allowable loads, which are:

- waste (effluent) load allocation (WLA) from major point sources;
- diffuse source loads allowable (DLA);
- loads allowable from background sources (BLA).

It is good practice to build some additional capacity into the management of the water resource (allowing scope for new developments or changed circumstances) by allowing a safety margin (allowable load margin of safety –MOS). Thus the total mean daily load (TMDL) is

$$\text{TMDL} = \text{WLA} + \text{DLA} + \text{BLA} = \text{LC} - \text{MOS}$$

The DLA – allowable diffuse load – in the above equation can be estimated by a range of techniques depending on the availability of data, monitoring resources and timescale for decisions (see Jenkins *et al.*, 2000). The concepts and methodologies for estimating the diffuse pollution loads range from simple annual unit loads appropriate for some lake management watershed plans to use of complex dynamic watershed simulation models.

BMPs are by definition techniques based on known levels of effectiveness; nevertheless, in the years since the concept was developed, much monitoring work has been done and there have been major advances in modelling techniques to enable better predictions of pollutant loads arising from various land uses, and of the effectiveness of BMP measures to reduce them. Jenkins *et al.* (2000) make a case for a crude screening exercise to be undertaken in respect of diffuse pollution risk assessments, prior to committing resource to complex models that require a lot of expensive data for validation.

Efforts to measure the effectiveness of point source discharges from diffuse pollution control structures such as ponds are relatively straightforward, given the need for storm event sampling and/or time of travel studies to characterize the performance of the facility. For landscape measures such as in-field practices and even linear measures such as buffer strips, catchment assessments are necessary. The USEPA has produced guidance on monitoring for determining the effectiveness of controls (USEPA, 1997a). That report comprises useful sections on: an overview of the pollution problem; developing a monitoring plan; biological monitoring; data analysis; quality assurance and quality control. An earlier, but still useful, document is also from the USA (US Department of Agriculture, 1994) on evaluating the effectiveness of forestry BMPs. Examples of the effectiveness of urban and rural BMPS are given in Chapters 4 and 6.

The Water Framework Directive in Europe and TMDL plans in the US are essentially watershed management plans. To avoid mistakes of the past, these plans now must have a mandatory and feasible/enforceable implementation component and time schedule of implementation. Sources of financing and who pays for the implementation must be identified and agreed upon among the agencies and stakeholders. Role and participation of stakeholders must be included from the beginnng of the planning process. The plan must also address the post-implementation management and continuum of the process (Committee to Assess the Scientific Basis of the TMDL Process, 2001).

Poop scoop bin, Malmö, Sweden.

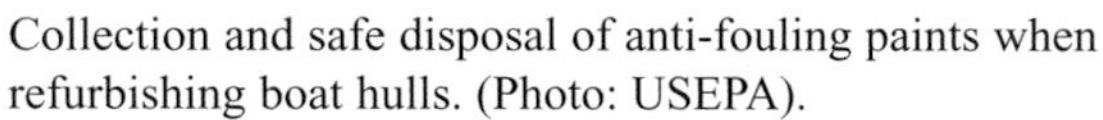

Collection and safe disposal of anti-fouling paints when refurbishing boat hulls. (Photo: USEPA).

Silt interception and sedimentation can minimize pollution during the construction phase of development, Maryland.

Waste oil collection facilities at a marina in USA (Photo: USEPA).

Conservation tillage encompasses a range of cultivation techniques to minimize soil losses, Wisconsin.

Pesticide mixing and rinsing pads for sprayers avoid contamination of soil and water, if isolated from drainage systems and groundwater,Wisconsin. Biobeds may effectively treat run-off.

Buffer zones of undisturbed native vegetation along watercourses are widely used in planning forestry plantations: River Tweed Catchment (right) and the Kielder Forest (above).

3

An introduction to BMPs for the built environment

3.1 GENERAL PRINCIPLES FOR URBAN SURFACE WATER DRAINAGE MANAGEMENT

Surface water drainage management and source control techniques, globally referred to as best management practices (BMPs) have been developed and successfully used in USA and parts of Europe for a considerable number of years to effectively control surface water run-off quality and peak run-off flows to prevent flooding and enhance water quality (Urbonas and Stahre, 1993). Urbanization of natural greenfield and agricultural land has significant effects on the quantity and quality of surface water run-off generated, and these effects of urbanization are outlined briefly below. In practice it is difficult to separate hydrological impacts from those associated with diffuse pollution. An integrated approach to developing appropriate best practice technology is therefore necessary.

3.2 EFFECTS ON RUN-OFF

Depending on climate and soil type, relatively little (15–20 per cent) of the rainfall falling on undeveloped land appears as direct surface water run-off. Most of the rainfall naturally soaks into the topsoil. There it is either held by capillary action, or slowly migrates into and through the soil (as interflow) to the nearest watercourse or into the general groundwater. The remaining rainfall on the surface runs off the catchment slowly, so that rainfall effects are spread out over a period of several hours. Even short duration, high intensity rainfall events may have relatively little impact on peak flow rates in the receiving watercourses.

As a catchment develops, the land covered by impervious surfaces (roads, parking areas, roofs, driveways, pavements) increases, preventing the natural infiltration of rainfall into the topsoil and underlying ground substrata. Even the remaining open ground, or pervious surface, often cannot accept infiltration as rapidly as it did before as, during construction, topsoil is often removed or mixed with the underlying subsoils which may be less pervious. Perviousness is further reduced by conventional urban surfaces such as concrete, tarmacadam, and roofing tiles. In these circumstances with potential infiltration much reduced and impervious areas increased, the quantity of direct run-off can increase to over 80 per cent of the rainfall volume. At the same time, since water travels more rapidly over paved and impervious surfaces, the run-off rate is much more sensitive to rainfall intensity and volume.

Therefore, in general, volumes and rates of run-off both increase significantly after development. Peak flow rates can increase by a factor of two to ten times.

An example can be cited from the state of Maryland in the USA. There it was found that stream peak flow rates that were exceeded once every two years before development, would be exceeded three times per year if the area was developed as residential, and six times per year if developed as commercial property (Schueler, 1987). This effect is shown graphically in Figure 3.1.

The overall effect of urbanization is that the flow frequency curve for a given area is significantly higher for a developed area than for an undeveloped area. This will tend to be exacerbated, in those parts of the world where climate change/global warming is causing an increase in rainfall amount, intensity and frequency (as noted in recent years in the UK).

3.3 EFFECTS ON RECEIVING WATERS

The most common visible effects of urbanization on receiving waters are the evident physical degeneration and erosion of natural streams and their banks including the presence of litter, sediment and oil sheens on the water surface. Decreases in biodiversity and numbers of aquatic stream biota are also common, if less obvious.

The higher frequency of peak flows causes the stream to cut a deeper and wider channel, destroying the in-stream aquatic habitat. Eroded sediments are deposited downstream in slower moving reaches, destroying aquatic habitat in these areas by smothering the benthos, filling wetlands with sediment (Roesner, 1999a).

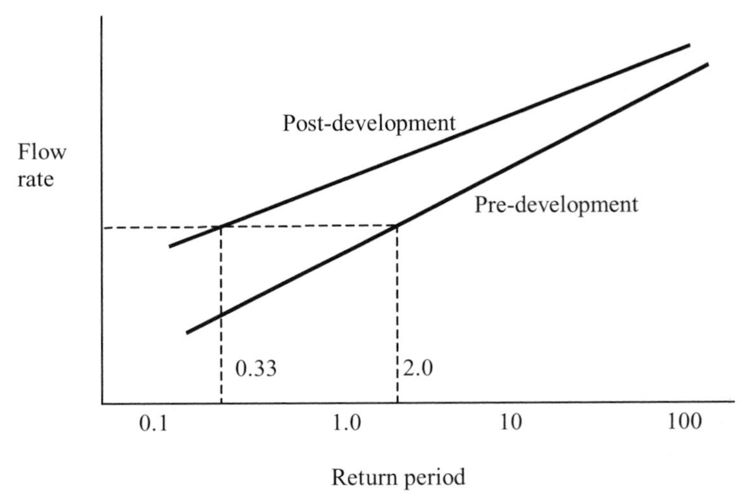

Figure 3.1 Effect of urbanization on peak flow discharge frequency (Roesner,1999a)

These hydrological impacts, associated with urbanization, are the reasons why so many urban rivers are protected in concrete channels, and streams are often culverted, resulting in little or no amenity or wildlife value, and, again, often increasing run-off rates with smoother channel beds and walls.

Oils and trace metals associated with motor vehicles and road run-off are generally regarded as toxic in increased concentrations, as are pesticides and oil–detergent mixtures. The more pollutants used in a catchment, the greater the risk of significant contamination of surface water. Urbanization in the form of industrial estates is generally regarded as being the worst-case scenario for surface water contamination, followed closely by that from roads and highways.

3.4 BASIC CONSIDERATIONS FOR SUSTAINABILITY

As previously explained urbanization has two main effects on post-development surface water run-off which require to be properly addressed. These relate to a substantially increased quantity and intensity of run-off and a significant reduction in run-off quality due to increased pollution. As well as adequately dealing with these two main effects of urbanization, the BMP control and treatment facilities provided for the development should ideally also satisfy the additional following basic considerations with regard to sustainability:

1 That the run-off from the developed area should not result in any downgrading of the downstream watercourses or habitat.

2 Urban stormwater run-off generated by urban development should be treated within the development area prior to discharge.
3 The urban drainage systems and techniques employed should be designed with consideration being given to water resource management and control within the development area and the environmental surroundings.

The first point is to secure environmental protection, the second to ensure that a development does not export problems it generates for off-site solutions at the expense of others. Finally, the last point requires that a holistic approach is taken considering opportunities for cost-effective solutions to drainage problems and that broader social and environmental considerations are taken into account (as appropriate to the scale of development). For example, the drainage solution chosen may be designed to also reduce quantities of stormwater in combined sewers, or to allow groundwater recharge by source control techniques, or simply to provide an amenity such as a multi-functional stormwater pond.

3.5 WATER QUALITY TREATMENT

The full range of urban BMPs including swales, filter strips and drains, extended detention basins, retention ponds, wetlands, porous and permeable pavements and surfaces which, along with soakaways, infiltration trenches and basins, can successfully be used either individually or in various combinations to provide the type and level of treatment required. Processes range from filtration or sedimentation, (suspended solids removal) to biological treatment (biodegradation and nutrient uptake).

It has been generally established for much of North America and Europe that the majority of rainfall falls in storms that are relatively small; it is these frequent but relatively minor rainfall events that mobilize pollutants on urban surfaces and transport them to watercourses (Roesner, 1999a). Furthermore, in the larger, less frequent storms, the greatest run-off pollutant level is found to be early in the storm during the initial period of rainfall and run-off which is generally referred to as being the 'first flush' (CIRIA, 2000). The levels of pollutants found in the run-off from the larger, less frequent storms are noted to reduce after the 'first flush' period of the storm in urban catchments, although the shape of the pollutograph varies considerably from storm to storm, place to place and between different pollutants. (see Deletic, 1998 and Yamada. *et al,* 2002).

Nevertheless, considering this it would seem to be reasonable and most efficient that urban BMP facilities, which are required to provide quality treatment to the development surface water run-off, need only treat the surface water run-off from the small, more frequent storms and the most polluted 'first flush' surface water run-off from the larger less frequent storms. This means that BMP facilities specifically and solely provided for water quality treatment can therefore be significantly smaller in size than those dealing with flood control. It also means that alternative, additional provision must be made for these larger flows, unless accommodated by modifying the BMP design.

The type of treatment sought from BMP facilities varies according to the pollution risks inherent in the types of development, and also with the overall provision of BMPs across the development (see Section 3.7 and Chapter 2). The most usual treatment process is sedimentation, but sometimes biological treatment is required. Text Box 3.1 clarifies common misunderstandings about biological treatment.

3.5.1 Water quality treatment volume

A methodology to determine the level of water quality treatment to be provided as defined by the specific hydrologic and climatologic data applicable to the development site/area has been developed in the USA (Roesner *et al.*, 2001). This methodology allows for a percentage of the annual rainfall captured for treatment to be determined that takes accounts for all small storm run-offs and the first flush run-off from the larger storms and translates this into the form of a basic treatment

Text Box 3.1 Biological treatment

Biological treatment of stormwater run-off can occur as the water flows slowly through a pond or wetland, or infiltrates slowly through surface material into soil or sub-soil. It refers to the degradation of pollutants present in solution or suspension, and in surface systems to the uptake of nutrients by vegetation or algae. Biological treatment therefore requires significant retention time for stormwater so that entrained pollutants can be broken down by microbiological processes.

Biodegradation processes also obviously occur in various components of many BMPs (not just in stormwater wetland or retention ponds) but this term includes the slow breakdown of, for example, residual oil films on granular fill in sub-soil infiltration/detention systems, or in topsoil in detention basins or swales, and also deposited organic sediment in topsoil in grass swales. The existence of such processes does not mean that the system as a whole can be said to provide 'biological treatment' if a substantial portion of the influent load passes through the BMP before degradation can occur.

Work in the USA has identified a residence time of 14–21 days for a retention pond to provide adequate time for bio-degradation and nutrient uptake. Such systems are appropriate as regional facilities for major urban developments, and also for the much smaller, higher risk catchments of many industrial units and farm steadings. Fourteen days is suggested as adequate for stormwater wetlands.

(Roesner *et al.*, 2001)

volume (V_t /unit area of the catchment). The computer model STORM has been developed in the USA to facilitate this type of percentage capture method of analysis to determine the appropriate level of water quality treatment to be provided (Hydrologic Engineering Center, 1976). An alternative approximate method of determining V_t relating it to the Wallingford procedures site location data within the UK has also been developed based on calculated site-specific STORM analysis.

Examples outlining the use of both the STORM computer model and the alternative approximate method of determining the basic unit area treatment volume V_t can be found towards the end of this chapter.

Once the value of V_t for the development site has been determined the total design treatment volume TV_t for the catchment or sub-catchment in question can be computed by multiplying V_t by the appropriate total catchment or sub-catchment area as outlined in the following equation.

Total design treatment volume equation

$$TV_t = V_t \times \text{Total Catchment Area} \qquad (3.1)$$

The value of the total design treatment volume TV_t when calculated is used to volumetrically size the urban BMP water quality treatment facility chosen to serve the catchment or sub-catchment (Camp Dresser & McKee *et al.*, 1993). The application of the total design treatment volume TV_t to some common types of BMP treatment facilities/controls is described below:

Issues relating to the treatment volume multipliers noted below are outlined in CIRIA, 2000; Roesner *et al.*, 2001 and Roesner 1999b:

a) *Extended detention* The detention basin volume shall be equal to the total design treatment volume, TV_t computed from Equation 3.1. The outlet works shall be designed to drain the full basin within 24 hours but on-site basins serving commercial or small industrial sites draining to a regional treatment facility can have a reduced drain-down time of 12 hours. Detention storage for drainage peak flow control can be incorporated into/above the facility with the treatment volume being considered as detention storage volume in the drainage computations. A permanent pool up to 0.5 metres deep with perimeter emergent aquatic vegetation may be added for aesthetic reasons, but this permanent pool will however not be considered to be part of the treatment volume. Design multi-staged outlet to meet peak flow control requirements for various design storms. A *detention pond* has a permanent pool volume of TV_E.

b) *Retention ponds* The permanent pool volume of the retention pond shall be $4 \times TV_t$, where TV_t is computed from Equation 3.1. As a guideline, average pool depth should be 1.3–2.0 metres. The maximum depth should be limited to three metres. Emergent aquatic vegetation to facilitate biological treatment of the run-off should occupy 25 to 50 per cent of the pond surface area, concentrated particularly around the perimeter. Detention storage for peak flow control can be added, but should be limited to two metres depth above the permanent pool to prevent damage to the vegetation.

c) *Stormwater wetlands* The permanent pool for wetlands shall be 3 × TV_t, where TV_t is computed from Equation 3.1. The typical depth of the wetland should be 0.5–0.75 metres, (although deeper pools can be provided to achieve retention time and reduce surface area). A wetlands ecologist familiar with local wetland species should be consulted to determine the actual water depth. Emergent aquatic vegetation to facilitate biological treatment of the run-off should occupy 75 to 100 per cent of the wetland surface area concentrated around the perimeter. The depth of the open water areas must not exceed two metres. Peak flow detention storage control may be added to the permanent pool, but should be limited to two metres depth to prevent damage to the vegetation.

d) *Swales* designed for use as a treatment control shall contain the total design treatment volume TV_t computed from Equation 3.1 within the swale subject to the following constraints: maximum water depth to be 0.1 metres, the velocity in the channel does not exceed 0.3 m/s, and the maximum channel side slope is 1v: 4h (one step vertical to four steps horizontal). Ideally the longitudinal bottom slope of the swale should be kept at a shallow slope of no greater than two per cent, although overall longitudinal slopes of up to six per cent can be accommodated by the introduction of check dam structures spaced at intervals along the length of the swale to maintain the hydraulic head at the effective gradient of two per cent or less. The swale, above the cross-section reserved for treatment, may be used for drainage peak flow control, subject to the following constraints: maximum side slopes of the channel are 1v: 4h, and the maximum velocity when the channel is flowing full is less than 0.6 m/s to prevent scouring. Swale channels should ideally be sited alongside the area which feeds them and run parallel to it, so that the run-off from the impervious area sheet flows down the sides of the swale.

e) *Infiltration controls* may be used for treatment if the infiltration facility is situated on soil classes 1 or 2 (see Text Box 3.2 for a description of WRAP soil classes), and it can be demonstrated with field data that the proposed control will completely infiltrate the total design treatment volume TV_t computed from Equation 3.1, within 24 hours under average winter rainfall and soil conditions.

For b) and c), if no detention storage for peak flow control is to be added above the permanent pool level, the outlet works may be simple, but should provide a smooth stage-discharge relationship for discharge of stormflows from the pond, e.g. a V-notch weir transitioning into a sharp-crested horizontal weir.

3.6 FLOOD CONTROL AND WATER RESOURCE MANAGEMENT

Methods presently used to alleviate potential flooding problems associated with increased urbanization tend to include measures such as the upsizing/enlargement of drains and sewers, the installation of detention/attenuation tanks within the developments piped drainage system and, occasionally, the use of larger downstream flood balancing ponds.

The main aim of conventional piped surface-water drainage systems is to convey the surface-water run-off from the impermeable areas within the urban development as quickly as possible. In terms of preventing localized flooding this would initially

Text Box 3.2 WRAP soil classes

Within the *Design and Analysis of Urban Storm Drainage* (The Wallingford Procedure, 1981) the types of soils found in the UK are identified and classified in general permeable or impermeable terms into five distinct classes by their winter rain acceptance potential (WRAP).

The five WRA. soil classes along with a general description of the relevant soil types are outlined below:

1. Well-drained permeable sandy or loamy soils and shallower analogues over highly permeable limestone, chalk, sandstone or related drifts.
   Earthy peat soils drained by dikes and pumps.
   Less permeable loamy over clayey soils on plateaux adjacent to very permeable soils in valleys.
2. Very permeable soils with shallow groundwater.
   Permeable soils over rock or fragipan, commonly on slopes in western Britain associated with smaller areas of less permeable wet soils.
   Moderately permeable soils, some with slowly permeable subsoils.
3. Relatively impermeable soils in boulder and sedimentary clays, and in alluvium, especially in eastern England.
   Permeable soils with shallow groundwater in low-lying areas.
   Mixed areas of permeable and impermeable soils, in approximately equal proportions.
4. Clayey or loamy over clayey soils with an impermeable layer at shallow depth.
5. Soils of the wet uplands with peaty or humose surface horizons and impermeable layers at shallow depth, or deep raw peat associated with gentle upland slopes or basin sites, or bare rock cliffs and screes and shallow permeable rocky soils on steep slopes.

appear to be beneficial, however the significantly increased peak flow rates generated in the receiving waters in using traditional piped surface-water drainage with little or no flow control could potentially cause serious erosion and flooding problems downstream as well as in-stream ecological damage.

The use of urban BMP surface-water management techniques, which can provide significant natural storage for the increased post development surface-water run-off, i.e. detention basins, retention ponds and wetlands, can have considerable benefits in terms of flood control. The natural attenuation storage provided by employing these types of BMP techniques should prevent flooding both within the development area and downstream of it to a marked degree (Larsson and Karppa, 1999).

As a general principle, the minimum levels of attenuation that should be provided within any development should be such that the post-development peak rate of run-off from the development area should not exceed the pre-development (greenfield) rate of run-off from the proposed site. This minimum provision of attenuation should, hopefully, ensure that the downstream receiving waters and their habitats are not damaged or downgraded to any significant degree and that any downstream flooding is not worsened by the level of controlled development run-off generated.

3.7 THE SURFACE WATER MANAGEMENT TRAIN

Controlling run-off quantity and quality should be in the forefront of the mind of urban planners and drainage designers from the initial development planning through to the final design. The surface water management train (D'Arcy and Roesner, 1999) is a useful concept to keep in mind during the development and design processes. The idea of the surface water management train is that the control of run-off quantity and quality is addressed at all stages of the development and design processes, making the final decision on appropriate BMPs easier and generally more cost effective.

The surface water management train is shown in Figure 3.2 with the general outlines of its four main consideration items

3.7.1 Good housekeeping

Good housekeeping measures are good practice employed to reduce the risk of chemicals and other potential pollutants coming into contact with the surface water run-off. Examples are given in Chapter 2.

Educational material and programmes which increase public and business awareness of potential inappropriate pollution problems should be developed and encouraged by all regulatory bodies, the local authorities, and the water and sewerage authorities.

Effective good housekeeping measures are pre-requisites for optimizing effectiveness (and amenity values) of BMP systems that are designed to deal with the unavoidable residual levels of contamination.

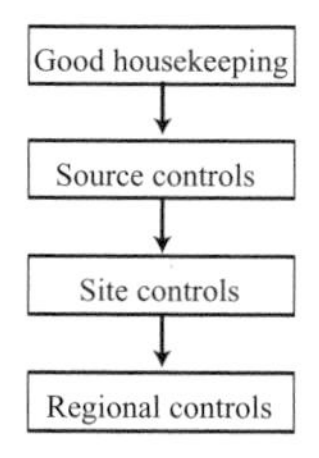

Figure 3.2 The surface water management train

3.7.2 Surface water source controls

Source controls attenuate run-off and help to remove pollutants from the direct run-off at, or near to, the source. These controls, generally serve small single-plot or sub-plot drainage catchment areas of typically up to two hectares, which is generally taken as the maximum drainage catchment area where direct run-off can practically be dealt with at or near its source. They can include simple measures, such as directing roof, driveway, and footpath run-off over grassed areas where, if possible, the run-off has an opportunity to infiltrate, or sheet flow through grassed areas, slowing the rate of run-off and allowing pollutants to settle out.

Source controls would generally include the following facilities and techniques:

- filter drains
- filter strips
- infiltration trenches
- simple measures directing run-off across grass (the MDCIA concept)
- porous/permeable pavement surfaces
- soakaways
- swales.

The type of measure where surface-water run-off at or near its source is directed across or through permeable or vegetated areas effectively minimizing directly connected impervious areas introduces a philosophy which is commonly referred to as the MDCIA (minimize directly connected impervious area) concept (Urbonas, 1999). The MDCIA concept is self-explanatory, but whilst practised in parts of Europe, the USA and by some individual property owners in Britain, it has not been widely accepted as an institutional practice in the UK, possibly due to the need to ensure that the current requirements of the Building Regulations are recognized and met by the proper design, which is not always possible. It does, however, offer potential benefits in both water quantity and water quality terms, not least in the elimination of risks of foul drains being wrongly connected into surface-water pipes (material from a foul sewer is instantly recognized as an error when appearing in a garden swale or blocking a gravel channel near the house). The consequences of the error are close to the source of the error, which is another benefit.

Gravelled or porous/permeable parking areas in car parks are effective and economic source controls. Roadside swales, filter strips, filter drains and, in areas that have soils with high infiltration capacity, soakaways and infiltration trenches also offer excellent source control.

It is very important that source control techniques are considered for all developments and used whenever possible, because the greater the amount of flow that can be controlled and treated near to source, the smaller the flows that need to be dealt with by larger downstream facilities.

In the cases where good soil porosity and permeability exist, and infiltration source control techniques and facilities can therefore be used, they should be able to effectively provide all the water quality treatment required without resorting to the use of either additional site or regional treatment controls. Other water quantity considerations and requirements may, however, necessitate the provision of some attenuation and/or flood routing measures.

3.7.3 Site controls

Site controls are run-off and treatment facilities that commonly serve site drainage catchment areas of two to five hectares and are used where the use of source controls to deal with the direct run-off at source has either been limited or is not possible for some reason.

The typical scale of development which site controls could serve would be a shopping centre, an individual, small industrial development or a residential development of 20 to 50 homes including their access roads.

Site controls deal with surface water run-off from the development which is collected and conveyed to the controls across the site by some form of conveyance system, and are therefore typically at the end of a conveyance swale or piped drain. Where a shallow swale with low longitudinal gradient is used as a conveyance system to a site control, the swale does provide some water quality treatment benefit and, as such, the sizing of the site control in water quality terms can be reduced commensurate with the level of water quality treatment, which the swale provides.

Site controls would generally include the following treatment facilities/ techniques:

- extended detention basins
- filter drains
- infiltration basins
- soakaways
- swales (in combination with a pre detention or filter basin).

As a general rule, where no biological treatment is required the most commonly used form of site control would be extended detention/attenuation basins, although infiltration when the soil type permits in the form of infiltration basins can be usefully employed as can porous/permeable pavement either with or without infiltration.

The use of retention ponds and wetlands for smaller development drainage areas do tend, from the sizing viewpoint, to be impractical and their possible use as site controls is therefore very marginal, but with careful design in sites with reasonable catchment topography they can be used for development drainage areas of four to five hectares or greater (but see Section 3.9 for rural situations).

If nutrient removal is not a high priority, the best performance for water quality will be extended detention with a permanent pool or wetland bottom. The provision of a permanent water pool or wetland area planted with aquatic vegetation in the bottom of an extended detention basin will significantly improve its aesthetic amenity value, increase the wildlife habitat while giving some potential for biological degradation.

3.7.4 Regional controls

Regional controls generally serve multi-hectare development drainage catchment areas greater than five hectares and when used in a fully integrated manner with upstream source and/or site controls they can provide the most comprehensive and optimum sustainable urban drainage control and treatment systems.

Regional controls would generally include the following treatment facilities:

- retention ponds
- stormwater wetlands
- enhanced extended detention basins.

As well as providing suspended sediment removal and attenuation/flow control retention ponds and wetlands can provide a level of biological treatment and nutrient removal. For some large-scale developments, where a treatment train of BMPs can be provided and pollution loads can reasonably be judged to be minimal, it may be that enhanced extended detention basins could provide an acceptable level of treatment as a regional control (but see Section 3.7.5).

When considering larger catchment/development areas, in which a number of separate sites are to be developed after the regional control has been constructed and established, it is essential to prevent the regional facility from becoming either hydraulically or biologically overloaded as development proceeds. To guard against this risk, a set of conditions must be included in the local authority planning conditions and consent, that clearly defines the maximum allowable fraction of impervious area in upstream developments on which the regional control has been

Text Box 3.3 Stormwater ponds: avoiding confusion

The technical literature includes references to a variety of terms for a range of different structures, all of which may be called 'ponds'. The pollutant removal and/or flood control functions of a facility are functions of specific design criteria, so it is very important to be absolutely clear what the purpose of a proposed pond is, and hence what sort of pond is provided.

Extended detention basins are sometimes referred to as 'dry ponds' since they drain down within 24–28 hours. A simple flood control detention basin may have a much shorter detention time (a few hours); the extended period (1–2 days) is required for any water quality benefit (principally the sedimentation of pollutants carried in suspension). Enhanced extended detention basins have a small permanent pool that helps prevent resuspension of sediment at the end of the drain-down period, and enhances amenity. Most ponds serving smaller developments are of this latter type. *Detention ponds* detain water for one week in the water body.

Infiltration basins are also sometimes referred to as 'dry ponds' since they are only flooded with stormwater for as long as the soil porosity and water table dictate for dispersal into the ground. Pollutant degradation and attenuation is in the topsoil of the basin, and to a lesser extent in the unsaturated zone around the facility.

Stormwater wetlands have a long residence time, and share the pollutant removal characteristics of retention ponds, differing only in the extent of the shallow water zone with emergent vegetation (75–100 per cent cover). The vegetation provides attachments for epiphytic micoflora and also dampens the effect of crosswinds that can stir up sediment in the open water ponds.

Retention ponds have a sufficiently long retention time (14–21 days) to permit degradation of pollutants and nutrient uptake by algae and/or higher plants. Stormwater retention requires a large area of land-take: retention ponds are normally found as regional facilities. Exceptions are some industrial and farm situations where soluble pollutants predominate. The large water volume of retention ponds and wetlands provides dilution for influent flushes of pollutants and hence minimizes any aesthetic pollution impacts within the facility.

The term *'wet pond'* should be avoided, since it fails to differentiate between extended detention basins with permanent pools and retention ponds – totally different facilities.

sized and designed. This allowable figure for impervious area fraction can then be checked when the future development sites are submitted for planning approval.

3.7.5 Combination control systems

When source and/or site controls are used upstream in conjunction with a regional control, the size of the regional control can be reduced commensurate with the level of pre-treatment and control that the source and/or site controls provide. In a combined BMP control and treatment system utilizing source and/or site controls along with a regional control the levels of water quantity and water quality treatment provided by each control can be varied to suit the specific requirements including where necessary the provision of biological treatment.

Perhaps frustratingly for some developers and regulators, there can be no absolute rules on selection of site and regional controls in relation to the range of local conditions and types of development, nor absolute thresholds to separate, say, a source control or site control, or site and regional controls. The final drainage scheme must remain a matter for the developers and regulators, following published guidance as quoted above, but tailoring the scheme to local conditions and pollution risks. Combinations of measures will always provide the best pollution prevention, and source controls will always be a desirable first consideration. These issues are further considered in the following sections.

3.8 SELECTING THE MOST APPROPRIATE URBAN BMP TREATMENT/CONTROL FACILITY

In selecting the most appropriate urban best management practice treatment/control for use as part of the surface-water drainage system, six inter-related factors have to be considered in some detail to establish the information required, both relative to the site and the proposed development, so that a properly informed and balanced decision can be made. These factors will almost certainly have a bearing on each other and should therefore be considered together and not in isolation (Campbell, 1998). The factors to be considered along with their sub-headings are outlined in Text Box 3.4.

As stated above choosing the appropriate BMP treatment control depends to some degree on each of the six inter-related factors outlined above, but an initial review of the type of development proposed along with hydrological site factors such as drainage catchment area and soil type can often usefully reduce the available BMP options to a more manageable number for further detailed consideration.

In this regard the Basic BMP Control Selection flowchart, as outlined in Figure 3.3, is a useful tool to aid the initial review of BMP options as it allows the main hydrological and biological treatment factors to be considered.

Text Box 3.4 Factors to be considered in selecting the most appropriate BMP treatment/control facility

Regulatory authority constraints and requirements

- Local authority
  - Building regulations
  - Local plan and planning conditions
  - Highway and road standards
- Water and sewerage authority
- Environment authority/agency

The type of development site

- Industrial developments
- Major road and highway developments
- Commercial and institutional developments
- Residential developments

Hydrologic site considerations

- Type of soil
- Drainage catchment area
- Land slope and topography
- Rainfall data

Environmental considerations

- Receiving water uses
- Combined control facility use for quality treatment and flood control
- Integration into the local landscaping
- Enhanced biodiversity

Community considerations

- Health and safety
- Creation of aquatic and wildlife habitat
- Enhanced amenity and recreation value

Adoption

- Long-term maintenance

Figure 3.4, whilst not precluding site-specific detailed development of a drainage system from first principles, offers a more indicative guide to the types of BMPs appropriate for the nature of the risks inherent in industrial development. The need to be able to track incidents, as well as provide a level of treatment for unavoidable levels of contamination favours selection of open drainage networks such as swales (ideally), or alternatively detention basins as site controls. The probability of a wide range of pollutants in the industrialized catchment favours the provision of a

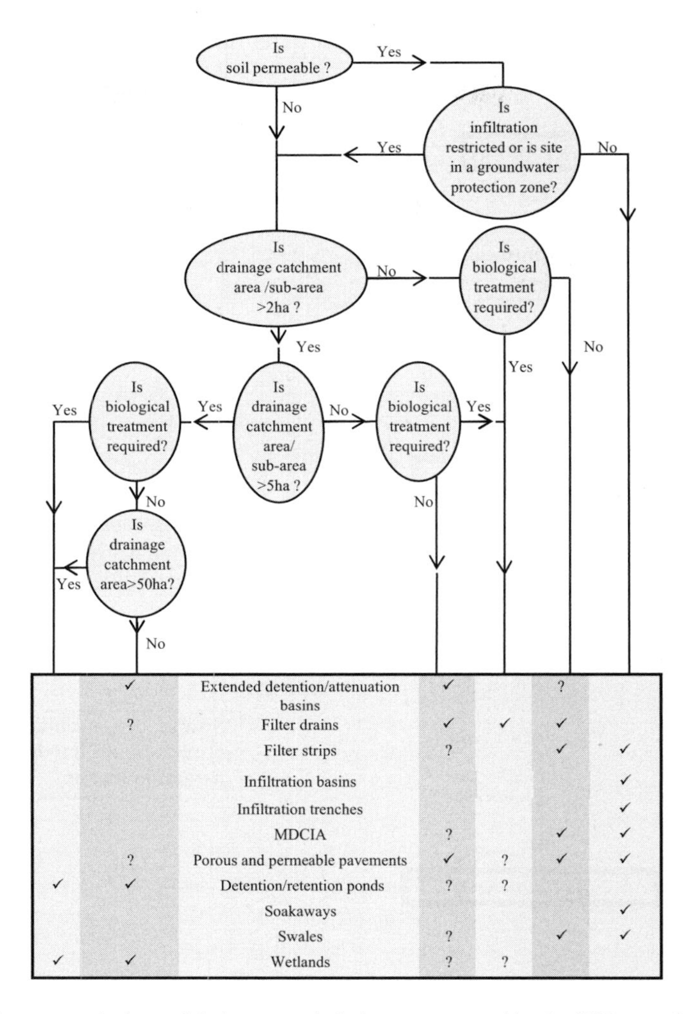

Notes:

- Catchment area is the total drainage or sub-drainage area served by the SUD control
- For smaller sub-catchments within a larger catchment, the SUD controls indicated on the table may be used, in which case the size of the SUD control to the larger catchment can be reduced
- Permeable soils means the soil is WRAP type 1or 2, and the depth to any impermeable layer or winter high water table is greater than 1 metre
- MDCIA = minimize directly connected impervious area

? Indicates application is questionable without careful design

Figure 3.3 Basic urban BMP control selection: hydrological and biological treatment considerations.

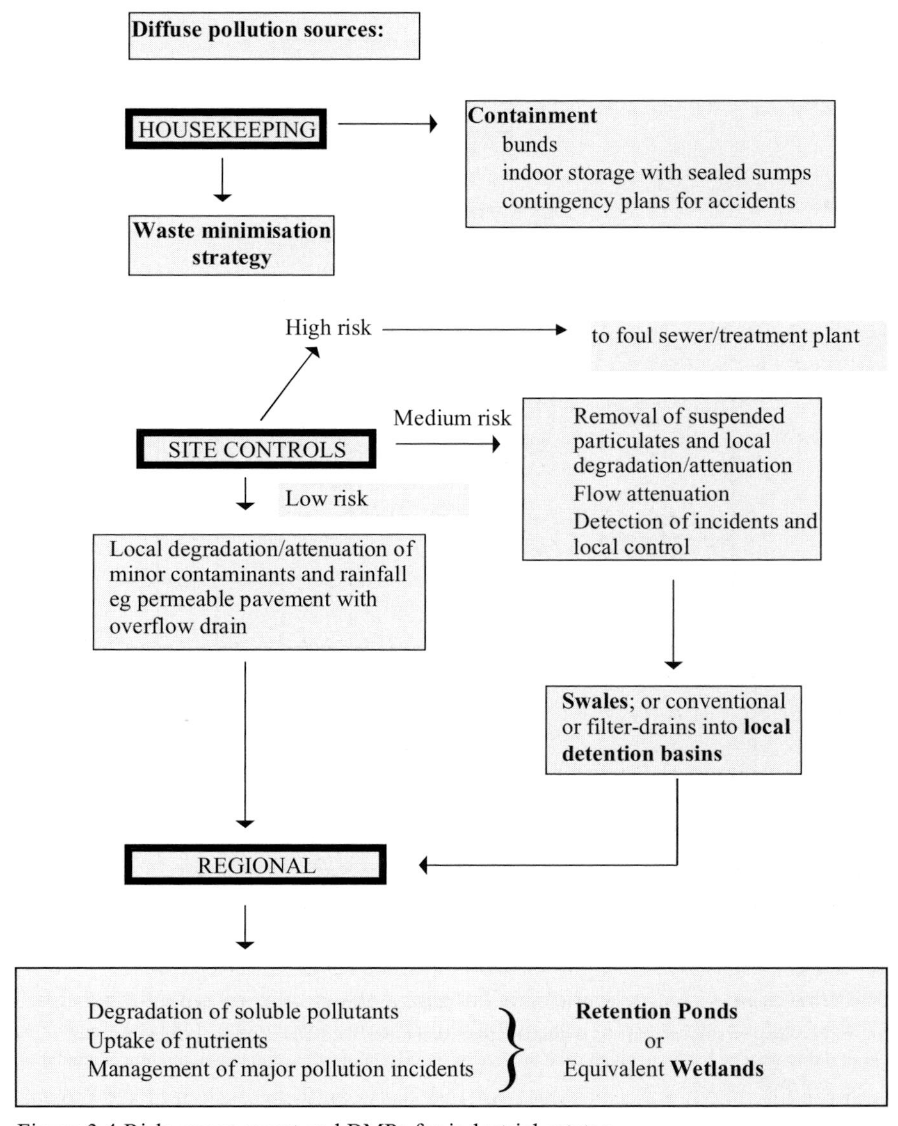

Figure 3.4 Risk management and BMPs for industrial estates.

regional control such as a stormwater wetland or retention basin, where soluble pollutants can be broken down and nutrients taken up. A large water body with quiescent flows is also a good risk management choice since such a facility allows for interception and recovery of oil spilled in major incidents.

Similarly, in Figure 3.5, it is evident that BMPs may be useful in combined sewer catchments, even where there is no watercourse available. Simply providing flow attenuation can have benefits in terms of reduced frequency of discharges from combined sewer overflows (Stovin and Swan, 2003).

In Figure 3.6 it is clear that BMPs can be part of the restoration of contaminated land, provided that infiltration systems are not used. Indeed, a swale offers the shallowest drainage option and hence least risk of puncturing the topsoil cover layer capping contaminated land. Where soil conditions require, the swale can be lined with an impervious membrane.

3.9 STRUCTURAL RURAL BMPS

The same water quality considerations that require BMPs for urban stormwater drainage – unavoidable levels of contamination and risks of spillages and accidents – also apply at the farmyard and other rural situations. The principal pollutants likely to be of concern are indicated in Table 3.1.

Swales alongside approach roads and farmyards will attenuate peak flows and achieve some removal of suspended matter. Alternative conveyance systems as used in urban areas for example, such as gravel filter drains, would probably require too much maintenance to be a sensible option for farmyard run-off, with its potential for carrying straw, glutinous muck, soil, spilled feed products and waste plastics into the drainage system. The importance of likely contaminants such as nutrients, faecal pathogens and intermittent traces of oil and pesticides in farmyard run-off indicates that retention ponds or stormwater wetlands are the appropriate site control for run-off from farmyards. The size of such farmyard drainage features can be minimized if swales rather than the conventional large drainage pipes can be used for some of the inflows. Example calculations are set out in Text Box 3.5.

Research in several countries has been undertaken to investigate the potential for artificial reed beds to treat farm drainage (Moir *et al.*, 2003; Harrington *et al.*, 2004).

Although much of the focus of that work has been the treatment of very strong farm effluents (e.g. slurry, silage liquor, dairy wash waters) there is also evidence that reed beds can be effective at removing much lower levels of contaminants (comparable to farmyard run-off), e.g. tertiary treatment reed beds used for polishing sewage effluents in the Netherlands and elsewhere, (Greiner and de Jong, 1984; International Office for Water, 2001) and for diffuse sources from fields (Braskerod, 2003).

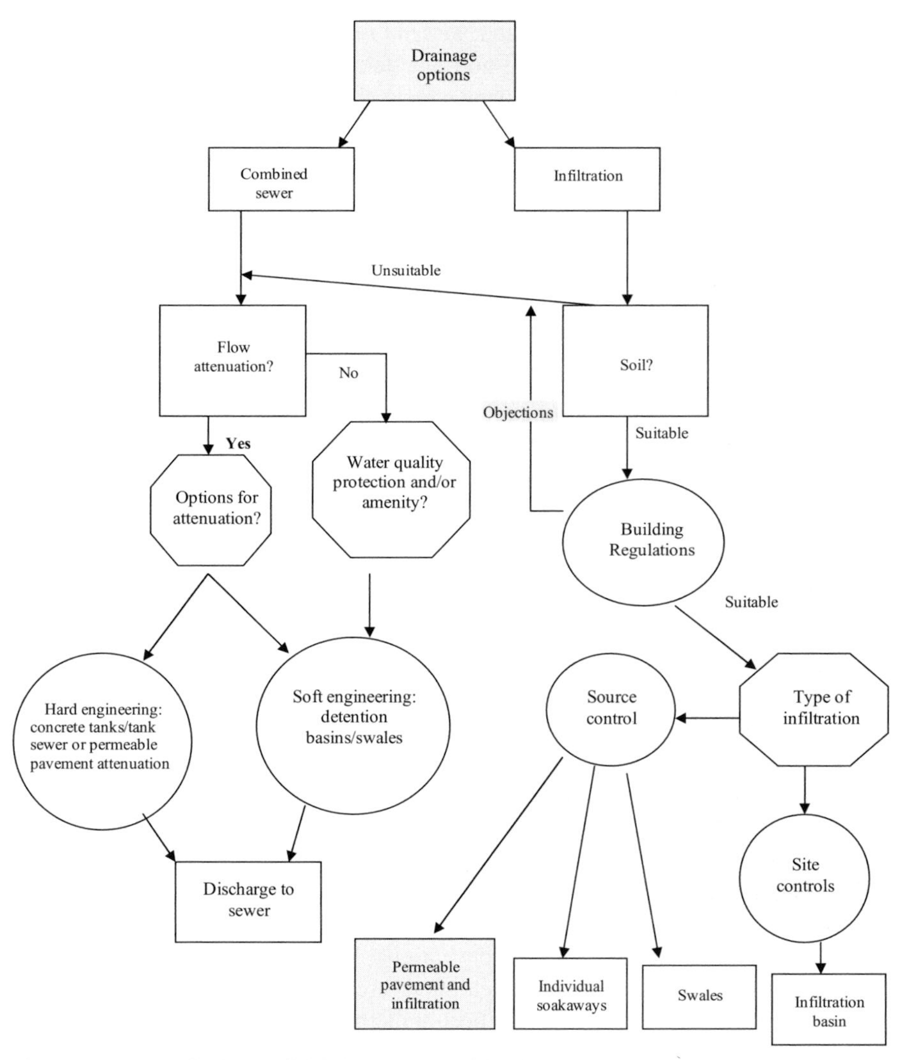

Figure 3.5 BMPs for brownfield sites on a combined sewer

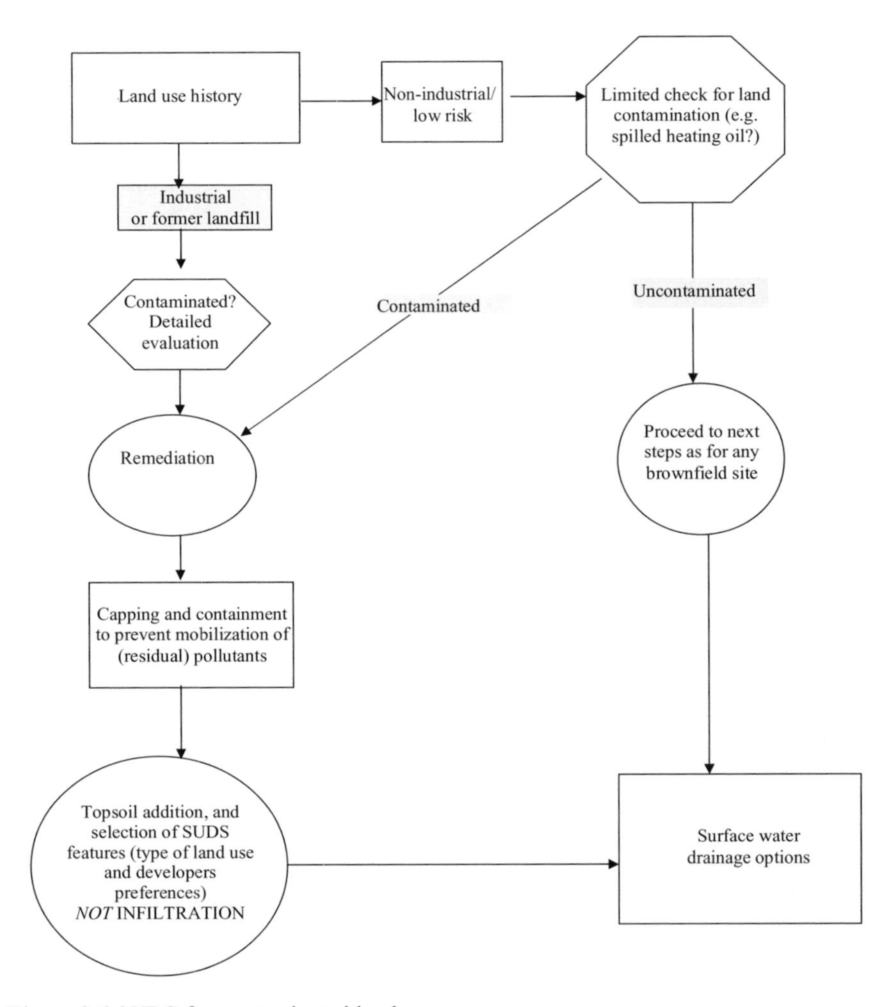

Figure 3.6 SUDS for contaminated land

Table 3.1 Pollutants from farmyards and other rural developments

Buildings/locations	*Possible contaminants in drainage*				
	Nutrients	*Faecal pathogens*	*Suspended solids*	*Pesticides*	*Oil*
Farmyard – arable	Y	N	Y	Y	Y
Farmyard – livestock	Y	Y	Y	?	Y
Poultry units	Y	Y	Y	Y	Y
Intensive pig units	Y	Y	Y	?	Y
Forestry depots/yards	?	N	Y	?	Y
Golf courses	Y	N	?	Y	?
Wherever sewage has to be disposed of in rural situations that will present an additional requirement for treatment (e.g. septic tank and reed bed or soakaway).					

Notes: Need for BMP facilities: Y indicates typically an issue, N the converse, and ? indicates a lot of variation between sites, or insufficient information.

Text Box 3.5 Sizing ponds or artificial wetlands for water quality treatment of farmyard run-off

If farmyard run-off treatment systems are to be widely adopted, it is essential that the cost be kept low. The first task is to separate, as far as possible, clean water from potentially dirty water. Clean water may include run-off from roofs and water entering the farmyard from fields upslope. Highly polluted water such as run-off from cattle feeding areas should also be separated and treated as slurry. The system outlined here is designed to treat water from lightly polluted surfaces only and also to act as a 'failsafe' in case of major accidents such as spillages of pesticides.

For a farmyard of 1ha total area with roofs accounting for 30 per cent of this and with an M5-60 rainfall of 13mm, a retention pond for water quality treatment would be approximately 350m^2 in surface area. Figure 3.7 shows a possible system.

For the same circumstances, a stormwater wetland for water quality treatment would require to be approximately 800m^2 in surface area.

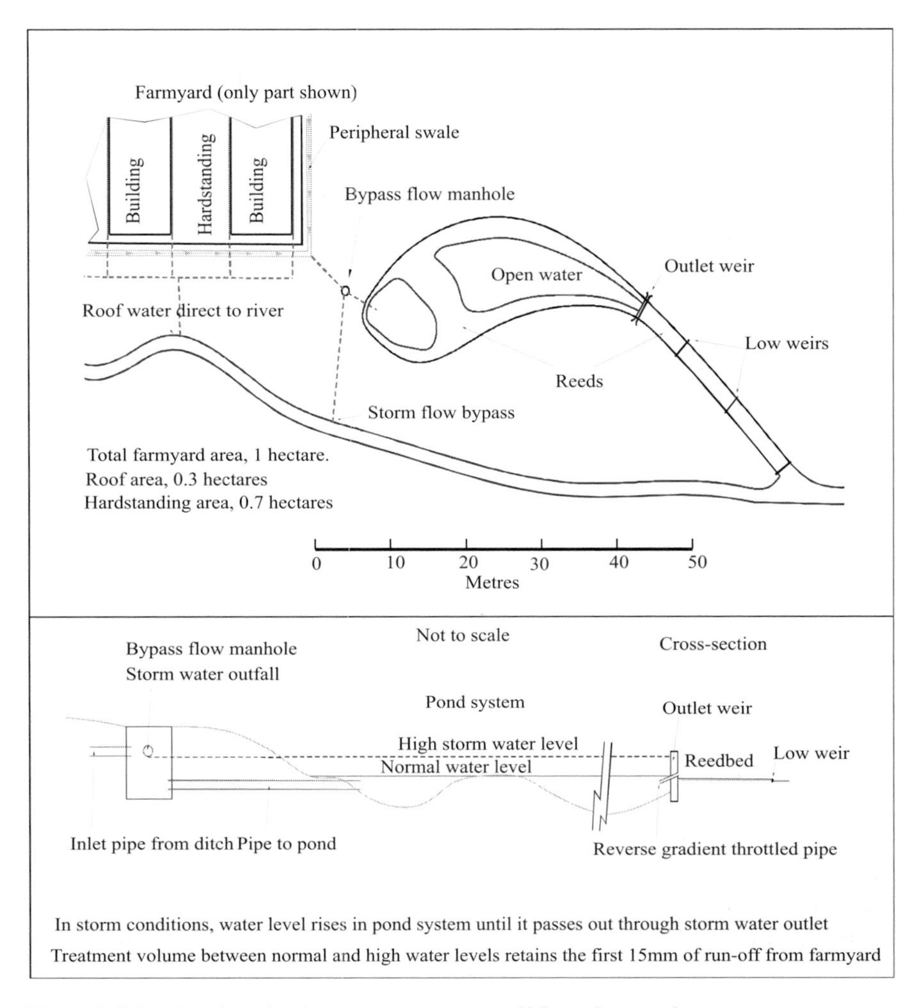

Figure 3.7 Pond and wetland system to treat run-off from farmyards.

3.10 AN INTEGRATED SUSTAINABLE APPROACH TO URBAN SURFACE WATER DRAINAGE

Run-off treatment and drainage facilities can be designed constructed and operated as single purpose facilities, which has been the practice in the past, or they can be integrated into the development plans in such a way that they serve multiple purposes. Treatment facilities can be designed as aesthetic community amenities and water features in the urban landscape, lending a sense of open space to an otherwise crowded development. Landscape areas and footpaths can be developed in the flood plain, providing the developer with more land to develop. But to be successful, a master plan must be drafted which has the support of all affected parties, i.e. the developer, municipal planning, highway and drainage staff, the water and environment authorities/agencies, and the affected public. If the project does not seek and gain this consensus, any one of the parties can cause the integrated plan to fail.

It is important to note that for many larger urban developments, the attenuation of surface water run-off within the development may be essential to prevent flooding. Water quality treatment considerations can be integrated into flood prevention facilities and equally flood prevention measures can be incorporated in water quality treatment facilities if both aspects are properly considered at the outset.

Urban development projects that integrate run-off treatment and flood control facilities have many advantages. The biggest is that the treatment facility can be incorporated into the flood control facility with little added space requirement, thereby increasing the amount of land that is available for development. Also, because the flood plain is rarely used, it can serve as a useful landscaping/amenity area, and the treatment facility can also be landscaped as a water amenity within the general urban landscaped area.

This holistic approach/philosophy, which results in the surface-water control facility providing combined water quality, water quantity and amenity can conveniently and readily be considered in the concept of the 'sustainable urban drainage triangle' as outlined below and in Figure 3.8. This triangular concept clearly shows the three main goals which this approach to urban drainage is attempting to achieve but it also gives each of these goals equal standing with the ideal solution providing all three.

When the functions of water quality treatment and water quantity control/flood prevention are combined in the same facility, there are important design considerations. The flood control portion of the facility should be designed so that during extreme events (two-year return storms or greater), inflows to the facility are slowed sufficiently so that they do not destroy the basin/pond habitat, or re-suspend pollutants settled out in previous storms.

These requirements are most easily met through the use of energy dissipation devices at the basin inlets and the use of multiple outlet structures which produce

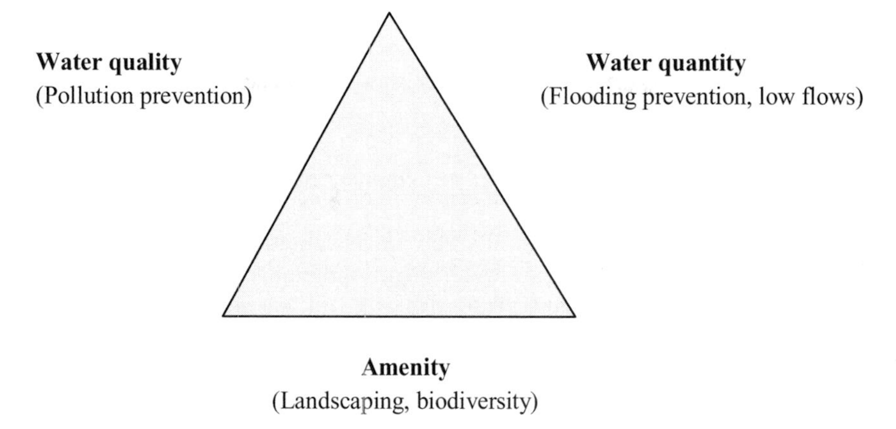

Figure 3.8 Sustainable urban drainage triangle.

reduced velocities through the facility at all levels. A multiple outflow structure provides a smoother outflow hydrograph from the facility over a wide range of storms, reducing channel erosion downstream and protecting habitat.

Run-off treatment facilities that include a permanent pool of water as part of the design or as an amenity should ideally included planted vegetation. The vegetation serves several purposes. In addition to improving the aesthetic appearance of the facility, it discourages humans from wading into the water; reduces wind-generated wave action, eliminating shoreline erosion and re-suspension of bottom sediments; and provides wildlife habitat.

Maintenance costs are likely to be minimized if the facility is designed to be self-sustaining with respect to wildlife and biota. Basic considerations include the use of plants that are appropriate for the local climate, and can thrive under the depth, frequency, and duration of inundation that they will experience in the facility; create habitat for desired aquatic life forms; and protect the facility from high velocities that will destroy habitat and vegetation.

It is generally considered that the integration of the objectives outlined above into the design and construction of BMP surface-water treatment and control facilities should not only result in these facilities performing their required water quantity and water quality functions well but should also provide the community within which they are built with an amenity in terms of aesthetics and wildlife habitat. This will be best accomplished by having the design carried out by an engineer experienced in this field, with input from an aquatic biologist/plant ecologist, and a landscape architect.

Although the benefits of an integrated approach combining water quality, water quantity and amenity have been emphasized above and this approach should be encouraged it should also be appreciated that this approach is not always a practical

Text Box 3.6 Designing for wildlife

Some key opportunities that every engineer should consider when designing stormwater management systems:

- Work with an ecologist and consider soft engineering opportunities, as well as hard engineering systems.
- Avoid culverting watercourses as culverts remove natural habitats, sever links between habitats and exacerbate water quality problems by preventing re-aeration of water and sediments. They also make tracing pollution incidents more difficult.
- When planning drainage schemes optimize connections between natural, semi-natural aquatic and marginal habitats, by working with landscape architects and ecologists to use swales and wetlands to form a continuous network with other greenspace areas, especially unmown riparian grassland, hedgerows and woodland.
- Grass swales can provide open corridors for wetland wildlife to colonize or recolonize habitats: benefits may be increased where landscaping requirements permit infrequent mowing – but designers need to allow for that by using check dams to provide a shallow aquatic zone across the flow of the swale. Otherwise tussock formation will result in a loss of the sheet flow that is usually achieved by a trapezoidal shape for a swale with short, regularly mown grass.
- The wildlife value of ponds is enhanced by achieving topographical variety – the shallow marginal shelf usually required for safety purposes offers an excellent opportunity for variety of width and depth; *avoid* over-engineering that achieves consistent width and level uniform surfaces.
- Peripheral or off-line areas at the edge of ponds and wetlands (by-passed by main storm flows) offer refuges for wildlife to recolonize ponds and wetlands in the event of serious pollution incidents, as well as providing landscape variety.
- Minimize planting: allow natural colonization as far as possible. Where planting is a requirement (e.g. for barrier zones and soil stabilization) use native species, locally sourced if possible, to avoid introduction of exotics or pests.
- Advocate and plan for a treatment train of BMPs, to ensure the site or regional features have the best possible water quality.

possibility particularly with smaller development sites where due to their size the choice of BMP treatment/control facilities is on occasion restricted to those where full integration is not always possible.

3.11 EXAMPLES DETERMINING AND USING THE UNIT DESIGN TREATMENT VOLUME

3.11.1 Using the STORM model to determine the unit design treatment volume V_t

The STORM model, as described above, was used to analyse four consecutive years of hourly rainfall data for a large development site to the east of Dunfermline, Scotland and the results for this development site are shown in Figure 3.9 in which the 90 per cent capture level is specifically lined and noted. This 90 per cent capture level is considered to be the optimum capture level as it results in the most cost-effective sizing of quality treatment volumes while still ensuring that the majority of the small storms along with the first flush fractions of the larger storms are captured and treated (Roesner *et al.*, 2001).

The 90 per cent capture level is therefore the level of capture which is to be adopted and used for the design sizing of surface-water quality/treatment facilities in this manual. The intersection points on the 90 per cent capture line on Figure 3.9 indicate that 90 per cent of the rainfall run-off can be captured and treated on an average annual basis by capturing the first 2.5 mm of run-off from the undeveloped naturally vegetated areas (classified as having an impervious fraction I = 0.0) and the first 11.5 mm of run-off from the developed surfaced areas (classified as having an impervious fraction I = 1.0).

It can also be seen from Figure 3.9 that the treatment/storage volumes relating to the 90 per cent capture level of the rainfall run-off varies nearly linearly with the impervious cover area of the development expressed as a percentage of the total area. This relationship between the percentage of effectively impervious area in a development to the treatment/storage volume is more clearly identified for the East Dunfermline development in Table 3.2 where the unit design treatment volumes V_{t},' for a unit development area of one hectare are noted against their relative impervious cover percentage taken from Figure 3.9.

This unit treatment volume V_t which is used in the design and sizing of a variety of BMP treatment control devices as outlined below, can also be calculated/defined for a catchment that contains both pervious and impervious cover as the area weighted average of the two volumes and is computed as follows:

$$V_t(\text{m}^3/\text{ha}) = 25 + (90 \times I) \quad (3.2)$$

where I is the fraction of the area that is impervious.

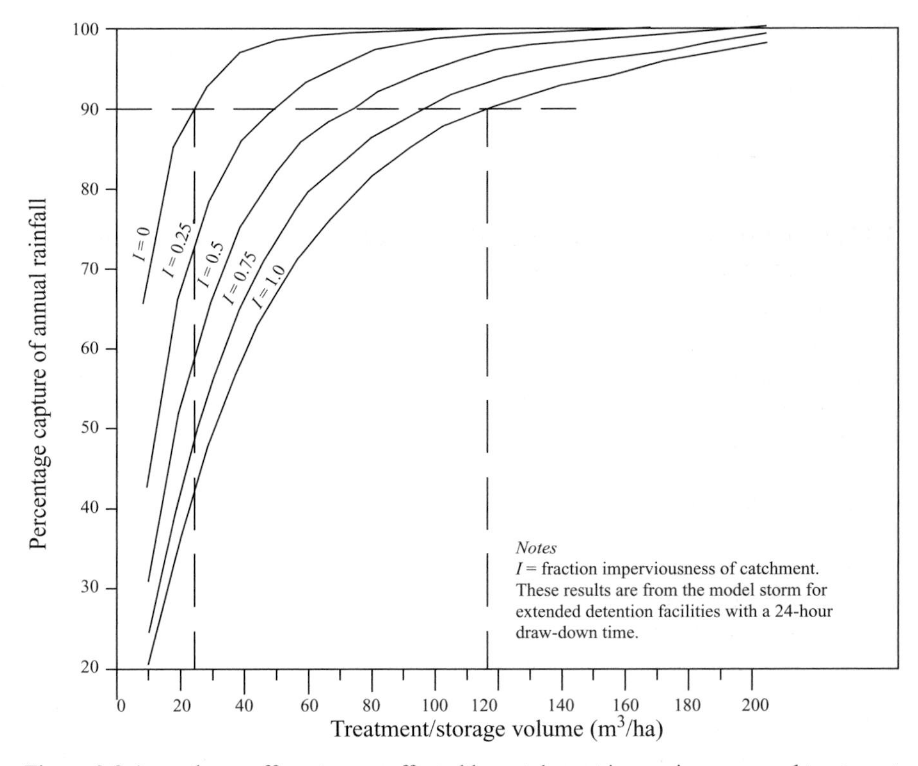

Figure 3.9 Annual run-off capture as affected by catchment imperviousness and treatment volume

Table 3.2 Unit design treatment volumes V_t

Impervious area (%)	*Volume (m^3/ha)*
0	25
10	34
20	43
30	52
40	61
50	70
60	79
70	88
80	97
90	106
100	115

Total design treatment volume TV_t

As noted earlier, to compute the total design treatment volume required for the treatment facility, multiply TV_t by the total catchment area in hectares that drains to it.

$$TV_t = V_t \times \text{Total Catchment Area} \tag{3.1}$$

3.11.2 Approximate method of determining the unit design treatment volume V_t in the UK

If extensive rainfall and hydrologic data to allow an in depth analysis is not available for the development site an approximation of the value for the unit design treatment volumes V_t likely to be obtained from the in-depth STORM analysis can be assessed using information on rainfall and soil classification contained within The Wallingford Procedure, (Department of the Environment, 1983).

Volume 3 of The Wallingford Procedure contains a variety of standard maps of the United Kingdom, which are used for the design and analysis of urban storm drainage. Two of the maps within Volume 3 are titled 'Rainfall depths (M5 – 60 minutes)' and 'Winter rain acceptance potential' and these maps and the information contained on them can be used to obtain a satisfactory approximation for the unit design treatment volume V_t for a development site where site-specific rainfall and hydrologic data is not available.

To determine the approximate unit design treatment volume V_t from the maps in Volume 3 of The Wallingford Procedure it is necessary to first establish with reasonable accuracy the location of the development site in question. Once this has been established the location can be plotted on both the M5 – 60 rainfall map and the WRAP map and the appropriate values for the M5 – 60 rainfall depth D in millimetres and the WRAP soil classification types 1, 2, 3, 4 or 5 can be determined. The soil index 'SOIL', as outlined in The Wallingford Procedure, has different values i.e. 0.15, 0.3, 0.4, 0.45 and 0.5 which relate directly to the WRAP soil classification types 1, 2, 3, 4 and 5, so when the WRAP soil classification type is determined from the WRAP map the appropriate value for the soil index 'Soil' can also be determined.

When the values for rainfall depth D and soil index 'Soil' for the development site are determined they should be put into Equation 3.3 along with the appropriate impervious area fraction I.

$$V_t(\text{m}^3/\text{h}a) = 9D(\text{Soil}/2 + (1 - \text{Soil}/2)I) \tag{3.3}$$

It should be noted that Equation 3.3 relates only to a 90 per cent capture level for water quality treatment which in the absence of any factual site rainfall and hydrologic data is generally considered to be the acceptable capture level for most cases.

Therefore taking the East Dunfermline development site as an example which has WRAP soil classification type 4 and therefore a value of soil index Soil = 0.45 along with a value of D = 13.4 mm. Then Equation 3.3 would be as follows:

$$
\begin{aligned}
V_t(\text{m}^3/\text{ha}) &= 9D(\text{Soil}/2 + (1 - \text{Soil}/2)I) \\
&= 9\times13.4(0.45/2 + (1 - 0.45/2)\, I) \\
&= 9\times13.4(0.225 + 0.775 \times I) \\
&= 120.6(0.225 + 0.775\times I) \\
&= 27.135 + 93.465I
\end{aligned}
$$

This equates reasonably well with the 25 + 90I value for V_t obtained from the in-depth STORM analysis of rainfall data for this site as outlined above with the value of V_t for this approximation method being, correctly, slightly more conservative than that obtained from the in-depth analysis.

3.11.3 Worked example using the approximate method for V_t

This example uses the approximate method to determine the design treatment volume V_t and then uses this value to establish the approximate master plan sizing of possible water quality treatment BMP facilities for a development site within the lowland belt of Scotland (residential, industrial and business uses).

Determination of the design treatment volume V_t using the approximate method employing rainfall depth and WRAP soil rating figures obtained from The Wallingford Procedure.

From the Wallingford M5 – 60 rainfall map the development area has a rainfall depth figure D of 14mm.

From the geological information obtained for the development area the surface and shallow drift deposits, which predominate are boulder clay and silts. These types of soils generally correspond to WRAP soil classification 4, which has an equivalent Wallingford Soil factor of 0.45.

Using the following approximation equation for the treatment volume V_t

$$
\begin{aligned}
V_t(\text{m}^3/\text{ha}) &= 9D\,(\text{Soil} / 2 + (1 - \text{Soil} / 2)\, I) \\
&= 9\times14\,(0.45 / 2 + (1 - 0.45 / 2)I) \\
&= 126\,(0.225 + 0.775I) \\
&= 29 + 98I
\end{aligned}
$$

where I is the impervious fraction of the area i.e. 40 per cent impervious I = 0.4

On a fully impervious surface such as a concrete hardstanding or tarred car park $I = 1.0$ and $V_t(\text{m}^3/\text{ha.}) = 29 + 98 \times 1.0 = 127\ \text{m}^3/\text{ha}$ which equates to a total rainfall depth of some 12.7 mm.

In comparative terms it is interesting to note that an equivalent one-year return twelve-hour storm for the development area would generate some 25 to 26 mm of rainfall depth, twice that required to be dealt with in purely water quality treatment terms. This gives a clear indication that facilities designed to deal with water quantity control will require to be much larger than those designed to deal solely with water quality treatment.

Sizing and selecting the appropriate BMP facilities

The basic quality treatment size requirement for BMP facilities in the development area has, as noted above, been determined as a 'treatment volume' related to a unit development area and can be calculated using the following equation:

$$V_t = 29 + 98.\text{I}\ \text{m}^3/\text{ha}$$

Where: V_t = treatment volume in m^3/ha of development
I = expected imperviousness of the development (expressed as a fraction)

For this preliminary BMP facility sizing exercise, it has been assumed that:

Residential development is 40% impervious so that

Residential $V_t = 29 + 98 \times 0.4 = 68\ \text{m}^3/\text{ha}$.

Commercial/industrial development is 60% impervious so that

Commercial/industrial $V_t = 29 + 98\ \text{x}\ 0.6 = 88\ \text{m}^3/\text{ha}$

'Park and ride' is 100% impervious so that

Park and ride $V_t = 29 + 98 \times 1.0 = 127\ \text{m}^3/\text{ha}$

For each case this volume of flow must be treated by a combination of source controls, followed by treatment controls that will improve the quality of the surface water run-off from the development area. The use of BMP facilities, which provide biological treatment, i.e. retention ponds or wetlands will be required for areas of 'park and ride' and industrial/commercial development either provided on-site or linked with an on-site pre-treatment extended detention basin if located off-site. For simplicity it has also been assumed that none of the treatment volume would be effectively restrained by the source controls provided upstream. Table 3.3 shows the volume of the treatment controls that would be applied to the development area can be defined as follows with the V_t multiplier, as noted in Section 3.5.1, relating directly to the residual treatment time required within each type of BMP facility

Table 3.3 Volume of treatment controls in different BMP facilities

	Housing	*Commercial/ industrial*	*Park and ride*
Extended detention	68 m^3/ha	88 m^3/ha	127 ^{3}/ha
Retention pond	272 m^3/ha	352 m^3/ha	508 m^3/ha
Wetland	204 m^3/ha	264 m^3/ha	381 m^3/ha

If we assume an average depth for each of these BMP facilities, we can then get an idea of what their surface area, shape and size on plan will be. The following average depths were used to compute surface area requirements: extended detention basins 0.5 m; retention ponds 1.25 m; and wetlands 0.4 m. These depths produce the planning level estimates of surface area required for the treatment control BMP facilities shown in Table 3.4.

Note that the basic size of the facilities identified above have been assessed only on the portion of the facility dedicated to water quality treatment control. In retention ponds and wetlands this relates directly to the pool of water, which permanently occupies these wet facilities. Additional attenuation storage volume for flood control should be added on top of this permanent treatment pool. However the addition of this flood control volume above the permanent water pool within these BMP facilities will generally not significantly increase the areas of these facilities as assessed using the sizing procedure noted above.

Residential Area (15.7 ha)

This residential development plot to the west of the development area is split into two approximately equal halves by the existing link road. There appears to be a distinct natural hollow in the land on the east boundary of this development plot adjacent to the north side of the link road. This would appear to be an ideal location for a treatment BMP in the form of a retention pond of 0.35 ha. or a wetland of 0.8 ha. with the outlet from the BMP facility being connected to a form of surface-water drainage conveyance network routed along the line of the main link roads. For a smaller development a detention pond may be suitable.

Table 3.4 Surface area estimates of treatment controls in different BMP facilities

	Housing	*Commercial/ industrial*	*Park and ride*
Extended detention	140 m^2/ha	175 m^2/ha	250 m^2/ha
Retention pond	220 m^2/ha	285 m^2/ha	400 m^2/ha
Wetland	510 m^2/ha	660 m^2/ha	950 m^2/ha

Class 4 Business area (11.9 ha) + Park and ride area (3.9 ha)

In terms of BMP quality treatment it would seem reasonable in this case to provide one BMP facility, which would serve both the 'park and ride' area and the business area.

The surface run-off from the 'park and ride' area could be collected and routed by means of swales to either a 0.5 ha retention pond or a 1.15 ha wetland located on the lower-lying north boundary of the business development area. The outlet from the BMP facility would then be connected by some from of drainage conveyance pipe or swale routed to connect into the main surface water drainage network on the line of the main link roads.

Industrial area (13.7 ha) + Park and ride area (4.0 ha)

This industrial area taken along with the adjacent 'park and ride' area could ideally be provided for in treatment terms by having swale drainage routing the surface water run-off into a 0.55 ha retention pond sited in the south-east corner of the industrial development plot. A 1.3 ha wetland could be provided as an alternative also sited in the south-east corner of the industrial area. The outlet from the BMP facility will be connected into the existing burn, which runs almost directly east along the south boundary of this development plot.

Northern class 4 business area (13.0 ha)

This business area is effectively split into two separate areas, with the west part being 3.7 ha and the larger east part being 9.3 ha. Each of these split parts will require treatment on-site using extended detention basins, 0.07 ha for the west sub-division area and 0.17 ha for the east sub-division. The discharge from these on-site extended detention basins should then be routed on to a 0.86 ha linear wetland sited along the northern edge of the development, with its outlet discharging to the existing stream, which runs along the north boundary of this development plot. An alternative to the linear wetland would be to provide a 0.37 ha retention pond.

3.11.4 Worked design example of an extended detention basin with wetland pool as an urban surface water quality treatment and surface water quantity control BMP facility

Site particulars

- Catchment site of 8.6 ha in Paisley, Scotland proposed for residential development.
- The site catchment has relatively shallow gradients between 2 per cent and 6 per

cent running down to a natural burn/stream, which bounds the site on its south side.

- The proposed residential development is assessed to have 45 per cent total impervious area associated with it.
- The impervious fraction I for this development site is therefore 0.45.

Surface water quality treatment design

Determine unit treatment volume V_t using the alternative approximate (Wallingford) method

Reference to The Wallingford Procedure Vol. 3 Maps would indicate that the M5 – 60 rainfall depth is approximately 16.6mm and the Winter Rain Acceptance Potential (WRAP) classification of the general soil on the site is Class 4.

$$V_t\,(\mathrm{m^3/ha}) = 9(\mathrm{Soil}/2)D + (1 - \mathrm{Soil}/2)DI) \qquad (3.3)$$

where from The Wallingford Procedure Vol. 1, Section 7.4, SOIL is the soil index and for WRAP Class 4 soil,

Soil = 0.45.
D = M5 – 60 rainfall depth = 16.6mm and
I = Impervious fraction = 0.45.

$$\begin{aligned} V_t\,(\mathrm{m^3/ha}) &= 9D(\mathrm{Soil}/2 + (1 - \mathrm{Soil}/2)I) \\ &= 9 \times 16.6(0.45/2 + (1 - 0.45/2)I) \\ &= 33.6 + 115.8I \end{aligned}$$

for $I = 0.45$

$$\begin{aligned} V_t\,(\mathrm{m^3/ha}) &= 33.6 + 115.8 \times 0.45 \\ &= 85.7\ \mathrm{m^3/ha} \end{aligned}$$

For site catchment area = 8.6 ha
Total design treatment volume TV_t

$$\begin{aligned} &= V_t\,\mathrm{m} \times \text{Total site catchment area} \qquad (3.1) \\ &= 8.6 \times 85.7 \\ &= 737.02\ \mathrm{m^3} \end{aligned}$$

Total design treatment volume TV_t

$$= \text{say } 750.0\ \mathrm{m^3}$$

Water Quality Treatment Discharge Control

(based on the normal 24-hour drain-down time)

$$750/24\times60\times60 = 8.7 \text{ l/s discharge}$$

Taking extended detention basin base as having a notional length (L) to breadth (B) ratio of 3:1 with the side slopes limited to a maximum slope of 1v: 4h.

Then basin treatment volume (V) in terms of basin base breadth (B) and water depth (D) would be

$$V = 3B^2D + 16BD^2 + 32D^3$$

Taking treatment water depth limit at say 1.0m, then

$$V = 3B^2 + 16B + 32$$

Equating V to total design treatment volume TV_t required of 750.0m^3, then

$$750 = 3B^2 + 16B + 32$$

or

$$3B^2 + 16B - 718 = 0$$

Solving for B gives

$$B = 13.04\text{, but take at say } 13.2\text{m}$$

With B = 13.2m,

$$L = 3B = 3\times13.2 = 39.6\text{m}$$

Extended detention basin base is approximately

$$13.2\text{m} \times 39.6\text{m} = 522.72\text{m}^2$$

With 1 in 4 side slopes and 1m water depth, the approximate treatment water area would therefore be

$$21.2\text{m} \times 47.6\text{m} = 1009.12\text{m}^2$$

Check actual treatment volume

$$V_{ta} = (522.72 + 1009.12)/2 \times 1$$
$$= 765.92\text{m}^3 > 750\text{m}^3$$

As discussed above an extended detention basin could be formed as a simple grass-covered depression with single or multiple inlets and an outlet structure which normally incorporates some form of granular/stone bound underdrain with a perforated outlet pipe leading to an outlet control set to limit the discharge to a required level but does not always including the required quality treatment discharge level, which is normally set to drain the full treatment volume in 24 hours.

This simple form of detention basin as noted above offers no great amenity value with little enhanced biodiversity or wildlife habitat being created and the underdrained outlet has been noted to be liable to unsightly silting up and/or blockage. The introduction of a wetland pool area at/near the outlet end of the detention basin has been found to have significant benefits.

Therefore provide a permanent wetland outlet pool with say 10 per cent additional volume set at an average depth of 0.25m below the extended detention base level and planted throughout with aquatic and marginal planting.

Sizing of wetland outlet pool

Additional 10% of treatment volume	= 0.1 × 750	= 75.0m^3
Average area of wetland pool	= 75.0/0.25	= 300.0m^2
Average breadth of wetland pool	= say 12.0/2	= 6.0m
Therefore average length of wetland	= 300/6	= 50.0m

Make wetland outlet pool approximately 50m × 12m max. set at and below the normal extended detention base level.

The sizing of the extended detention basin and its wetland outlet pool noted above is obviously based on a simplified rectangular pattern which should be modified as necessary to suit the natural contours within the area that the basin is to be located. A plan and details of this preferred form of extended detention basin with the wetland outlet pool extending naturally back up the basin towards the inlet is shown in the typical layout plan, and inlet and outlet detail drawings in Figures 3.10–3.13 at the end of this design example.

It should be noted that the extended detention basin noted above has been designed and sized to comply only with the surface water quality treatment criteria and as such it would not, as presently designed and sized, provide the level of quantity control likely to be required to deal with the increased post development surface water run-off from the site. Suitable forms of quantity control could be provided separately within the development site to satisfy the surface water quantity control criteria or the extended detention basin design could be amended to incorporate additional surface water attenuation storage above the calculated treatment volume which with the provision of multiple-stage outlet controls would allow the amended extended detention basin to provide both the required quality and quantity controls. The option of amending the extended detention basins design to incorporate additional surface water attenuation storage above the calculated treatment volume is outlined in the 'Surface-water quantity control design' which follows.

Surface-water quantity control design

On the basis that the proposed residential development is to be routed to a stream, which has a known downstream flooding history, then the local authority responsible for flooding would normally set the standard requirements in terms of attenuation and water quantity control to be provided to prevent increased post development downstream flooding.

Therefore presuming that the quantity control standards as set require that the post-development run-offs from the development site up to and including one in 100 year storm events are to be limited to a maximum predevelopment greenfield run-off value, set at 10 per cent of an M5 – 60 run-off, equivalent to approximately 5.25l/s/ha in this area.

Surface water run-off calculations

(based on a unit area (1.0ha) of development site)

The proposed residential development area is assumed to have an impermeable to permeable ratio of 45:55, which for 1.0ha of development site would equate to 0.45ha of impermeable hard surface and 0.55ha of permeable landscaping/garden surfaces.

Taking a basic run-off coefficient from the permeable surfaces/areas at say 10 per cent then the permeable areas of the development site can be accounted for in a 45:55 ratio residential development by taking 45% + 10% of 55% = 50.5% of the development area as effectively impermeable.

Therefore 0.505ha as fully impermeable for each 1.0ha of gross development area.

The peak surface water run-off flow figures and therefore the subsequent attenuation volume calculations will be calculated using The Wallingford Procedure – Design and Analysis of Urban Storm Drainage – Volume 4 – Modified Rational Method.

Therefore the basic data for use with the modified rational method is as follows:

- M5 – 60 rainfall = 16.6mm
- *r* ratio = approx. 0.25
- volumetric run-off coefficient *Cv* = 0.85

Determination of the required limiting 10 per cent pre-development peak discharge flow

For *r* = 0.25, Z1 factor for M5 – 60 = 1.00 ,

So from Table A2, Z2 ratio factor = 1.03, where Z1 and Z2 are Wallingford Procedure scaling factors.

The gross development site area at approx. 8.6ha in this case is relatively small for the areal reduction factor (ARL) to have any significant bearing on any calculated peak flows particularly when considering that the ARL would relate to both pre- and post-development calculations so take ARL = 1.0 in this case. Where Q_p is the peak discharge flow (in litres per second), then the required limiting pre-development peak discharge flow = 10% of Q_p for M5 – 60 storm where $Q_p = 3.61 C_v$AI, so pre-development peak discharge flow

$$= 0.361 C_v A I$$

where A = area of catchment (in ha) and C_v = volumetric run-off coefficient

Rainfall intensity I

$$= \text{M5} - 60 \times \text{Z1} \times \text{Z2} \times 60/D$$

$$= 16.6 \times 1.00 \times 1.03 \times 60/60$$

$$= 17.1\text{mm/hr}$$

For proposed residential development area limiting pre-development peak discharge flow

$$= 0.361 C_v A I$$

Limiting development area pre-development peak flow

$$= 0.361 \times 0.85 \times 1.0 \times 17.1$$

$$= 5.25 \text{ l/s/ha}$$

Therefore the post-development run-off from the proposed 8.6ha residential development should be limited to a maximum discharge of 8.6 × 5.25

$$= 45.15 \text{ l/s}$$

Post-development limiting discharge for water quantity/flooding control

$$= 45.0 \text{ l/s}$$

Determination of post development peak flows for the full duration range of 100 year return storms

As can be seen from the results in Table 3.5 above the maximum calculated attenuation volume, which needs to be provided to ensure that the required level of water quantity/flooding control is assured in the proposed residential development area is 217.14 m^3/ha.

So the total attenuation volume required to satisfy water quantity/flooding control requirements for the proposed 8.6ha residential development will be

Table 3.5 Post-development peak flows for M100 range of storms for equivalent net development area of 0.505ha.

Duration (minutes)	*Z1 factor*	*Z1.M5–60 (mm)*	*Z2 factor*	*M100 rainfall (mm)*	*Intensity (mm)*	*Post flow 3.61CvAI (l/s/ha)*	*Limiting pre flow (l/s/ha)*	*Net flow (l/s/ha)*	*Attenuation volume required (m^3/ha)*
30	0.76	12.62	1.97	24.85	49.71	77.05	5.25	71.80	129.24
60	1.00	16.60	1.98	32.67	32.67	50.64	5.25	45.39	163.30
120	1.28	21.25	1.96	41.65	20.82	32.28	5.25	27.03	194.58
240	1.66	27.56	1.88	51.81	12.95	20.07	5.25	14.82	213.44
360	1.95	32.37	1.83	59.24	9.87	15.30	5.25	10.05	217.14
600	2.32	38.51	1.80	69.32	6.93	10.74	5.25	5.49	197.81

$$8.6 \times 217.14 = 1867.4 \text{ m}^3.$$

Total attenuation storage volume for quantity/flooding control

$$= 1870 \text{ m}^3$$

This total attenuation storage volume will include the treatment volume previously calculated at 750 m^3.

Then total basin volume in terms of basin base breadth (B) and water depth (D) would be as noted above for the treatment volume case.

$$V = 3B^2D + 16BD^2 + 32D^3$$

Taking the basin base breadth (B) at 13.2 m as previously assessed, then

$$V = 523D + 211D^2 + 32D^3$$

Equating V to total attenuation storage volume required of 1870.0m^3, then

$$1870 = 523D + 211D^2 + 32D^3$$

Solving for D gives

$$D = 1.84\text{m}$$

for $V = 1876.0\text{m}^3$, but take D at say 1.85m.

As a purely quality treatment facility the extended detention basin's water depth to take the treatment volume V_t is approximately 1.0m with an outlet quality discharge control of 8.7l/s.

For the extended detention basin with a wetland pool to also act as a quantity/flooding control facility the basin's water depth rises to approximately 1.85m to take and store the required attenuation volume with dual outlet controls of 8.7l/s for quality discharge and 45l/s for quantity/flooding discharge. This combined outlet case is considered in more detail below.

The level of the extended detention basin with wetland pool emergency spillway, as shown Figure 3.10, would then be set at approximately 0.1m above either the treatment volume water level for a purely quality treatment facility or above the attenuation storage top water level for a combined quality treatment and quantity control facility.

The quantity/flooding outlet discharge control calculated at 45l/s, which regulates the total surface water run-off flow from the development would normally be provided by installing a flow control device, i.e. a hydrobrake, on the outlet to the receiving watercourse from the control manhole.

The quality treatment outlet discharge control assessed at 8.7l/s can be provided in a number of ways and three of these alternatives are listed below, which could be used in combination with a quantity/flooding outlet discharge control hydrobrake, as noted above, and be set in a dividing weir across the outlet control manhole, which is fitted with a penstock valve to allow the wetland pool to be drained down.

Three possible forms of quality treatment discharge control outlet, which can be fitted in a dividing weir wall and be designed to provide the required 8.7l/s maximum outflow are:

- V-notch weir outlet
- plate orifice outlet
- perforated riser outlet.

Outlet control manholes with each of these three treatment control outlets combined with a hydrobrake flow control are shown in Figures 3.11–3.13.

All three of the possible combined outlet arrangements noted above operate in a similar manner in that the treatment volume V_t within the extended detention basin

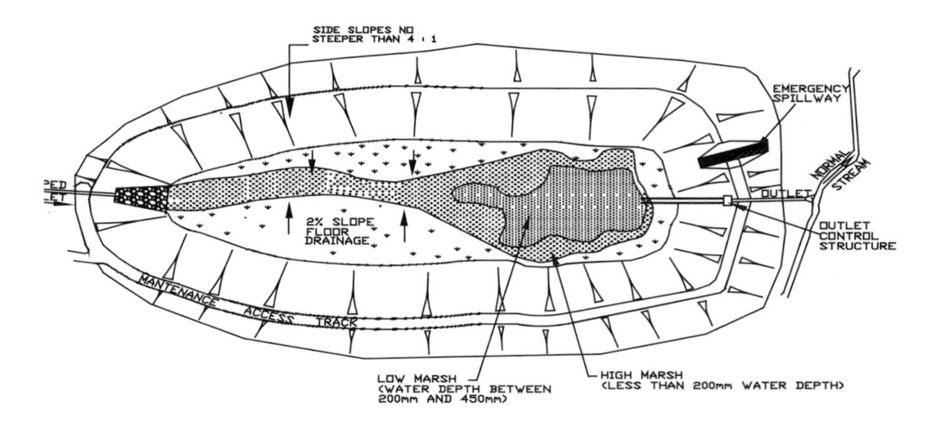

Figure 3.10 Typical layout plan.

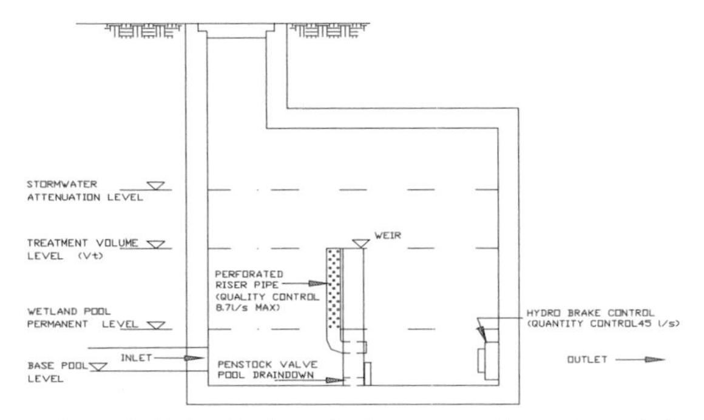

Figure 3.11 Typical manhole outlet with perforated riser.

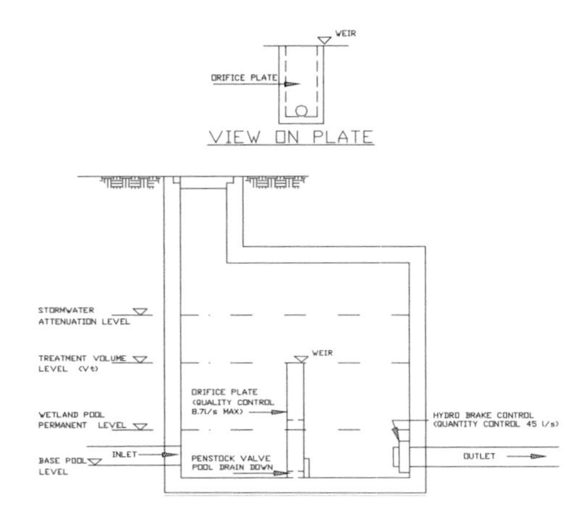

Figure 3.12 Typical manhole outlet with orifice plate.

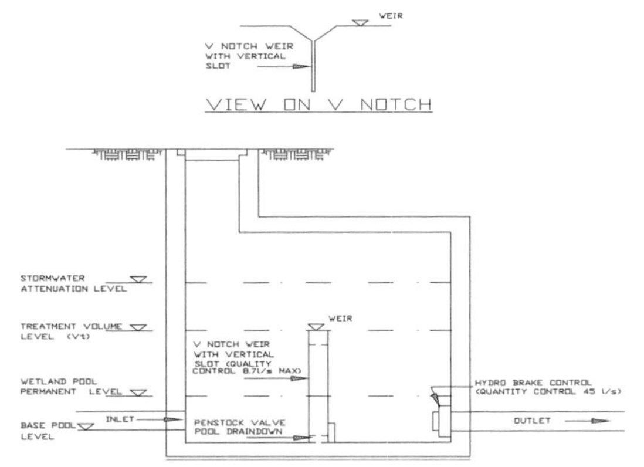

Figure 3.13 Typical manhole outlet with V-notch weir.

is held back by the dividing weir so that this volume of water must pass through the particular quality treatment discharge control outlet, which is fitted into the dividing weir wall and then flows through the quantity/flooding discharge control outlet (hydrobrake) to the receiving watercourse. This effectively means that the surface-water run-off from the majority of the smaller storm events and the first flush from the major storm events will pass through the quality treatment control outlet at the required discharge level. During larger storm events when surface-water run-off entering the extended detention basin exceeds the treatment volume it will overflow the weir and bypass the quality treatment control outlet going directly to and through the quantity/flooding discharge control outlet at a flow rate regulated by the flow control hydrobrake.

Above and right, swales in Wisconsin after wet weather storm.

Swale near Edinburgh, Scotland.

Swale in Malmö, Sweden.

Enhanced extended detention basin, Malmö Sweden.

Extended detention basin, Dunfermline, Scotland.

Extended detention basin, Dunfermline, Scotland.

Porous concrete car park, Orlando, Florida.

Dry permeable pavement near wet tarmacadam surface after rainfall, Scotland.

Infiltration basin in highway interchange, Orlando, Florida.

Retention pond serving industrial estate, Halmstad, Sweden.

Retention pond, Maryland.

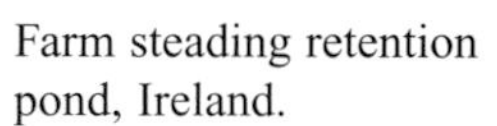

Farm steading retention pond, Ireland.

4

Managing diffuse pollution from urban sources: a survey of best management practice experience

4.1 INTRODUCTION: MAKING BEST PRACTICE ROUTINE

Do BMPs work and what evidence is available? What are the maintenance requirements for urban BMP facilities (ponds, permeable surfaces, swales, etc) and economic implications of adopting BMP measures? Before the USEPA launched its drive for the control of diffuse pollution in the mid 1990s, a great deal of data was collected to answer these concerns (USEPA, 1993). Much of the urban evidence available at that time is set out in the excellent publication by Schueler *et al.* (1992) which tabulates pollutant removal values reported for a variety of measures and in respect of a range of determinands, as well as usefully describing BMPs, their limitations, where they are appropriate and answering a number of basic questions. A great deal of research and monitoring has been undertaken in many countries since then and information from some of that work is set out in this chapter, using a case study approach and focusing on catchment scale improvements as well as individual local developments. The chapter considers the main issues in the implementation of urban BMPs for example situations primarily from the USA, Europe and Japan. As the number of monitoring projects proliferates, the range of values for performance data has expanded. The wide range of circumstances – appropriate and otherwise – where specific measures have been constructed is reflected in the variety of values reported in the literature.

4.1.1 Good technical guidance

There is considerable experience of urban best management practices (BMPs) in a growing number of parts of the world, but the concept is still unfamiliar in many places, or inadequately implemented. Uptake of some BMPs is sometimes inhibited by fears and uncertainties where individuals or institutions are unfamiliar with the approach or the technology.

Stormwater ponds are an example area where confusion about the design concept and appropriateness of the various options (see Section 2.4.2 and Text Box 3.3) has led to the provision of inappropriate facilities, unlikely to be able to meet the expectations of planners and regulators. It has not been uncommon for examples of 'failed' ponds – designed for stormwater flood control, but later assessed for their performance as pollution control facilities (BMPs) – to be seized upon as proof that BMPs don't work. An example study is Watt *et al.* (1999), a water quality appraisal of a flood-control stormwater pond in Canada that was not enlarged when its catchment area exceeded design specifications, and was never intended to achieve water quality improvements. Roesner (1999b) believes that much of the uncertainty about BMP performance, and part of the explanation for the wide range of pollutant removal performance reported, is due to inadequate specification at the outset of the stormwater facilities necessary to protect local receiving waters.

Good technical guidance is a pre-requisite for successful implementation of BMP technology. A wealth of excellent material is available in the USA (e.g. Schueler *et al.,* 1992; CDM, 1993; Horner *et al.,* 1994) and elsewhere (e.g. Pratt, 1999; CIRIA, 2000; Urbonas and Stahre, 1993) – but is it properly used and does it meet the needs of regulators, developers and others involved in providing urban infrastructure? McKissock *et al.* (2003a) reported on a project to evaluate technical guidance for urban BMPs used by developers, consulting engineers and others in the UK, together with awareness of diffuse pollution and SUDS (loosely equivalent to urban BMPs) terminology. Figure 4.1 shows what sources are used by engineers and others in Scotland. It reveals a reluctance to use material that is not local (Scottish in this study), reflecting the importance of soil conditions and hydrology, as well as institutional considerations where these are included in the manual. This is also

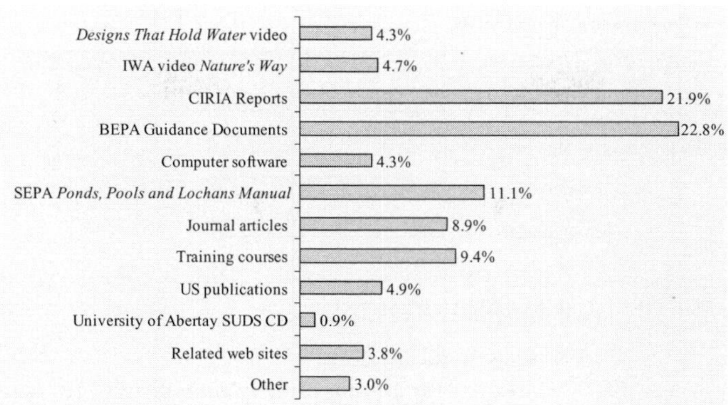

Figure 4.1 Usage of reference material

evident in production of series of state or city manuals in the USA (California, Maryland, Denver etc).

There was almost negligible use of websites or of slightly more difficult to access publications such as American textbooks and USEPA material in the Scottish study. That is surprising given the common elements within all the manuals (definitions of BMPs, construction advice, outlet design, the need to consider water quality storms as well as, or separately from, flood-risk storms and options to do that, and more). Other aspects of the survey showed the primary importance of design and construction guidance that is endorsed by the local regulators (McKissock *et al.*, 2003a). The survey also investigated barriers to considering alternative technology (Figure 4.2). These clearly go beyond technical considerations alone, and include institutional traditions and human behaviours too (see Text Box 4.1). Bray (2001b), in relation to maintenance for example (see Section 4.3 for more detail), concluded that the barriers associated with looking after SUDS elements (BMPs) in the landscape are more to do with a reluctance to change administrative conventions than with practical problems of site care.

Murray and Cave (1999) reported that a key finding of the Rouge River project (south-east Michigan, USA) was that major barriers to effective pollution control and water resources management are often not technical, but are institutional. The Rouge Project identified key elements needed in a regulatory framework to encourage participation in and acceptance of stormwater management by local governments. Catchment approaches (sub-watershed) are advocated to secure local interest in the issues. Target constituents included the general public as well as developers and public agencies.

The rest of this chapter addresses the main issues identified in Figure 4.2, supplemented by additional material from current or recent monitoring projects.

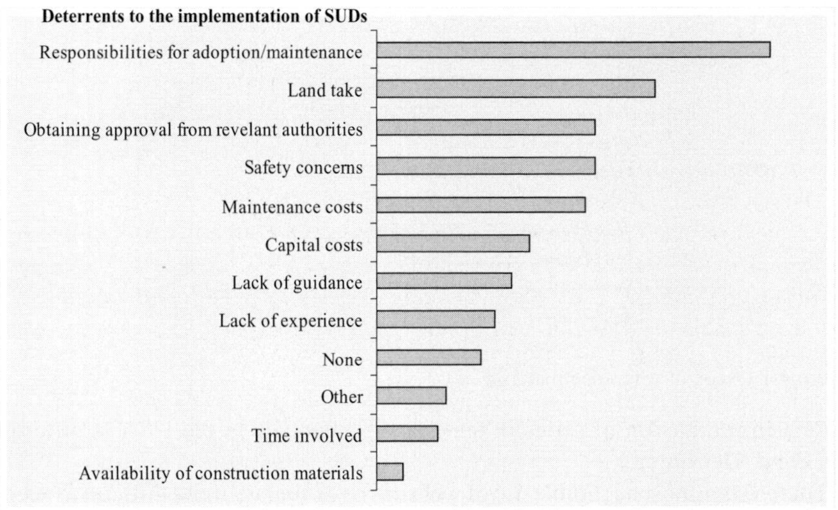

Figure 4.2 Barriers to provision of urban BMPs identified by questionnaire surveys (in Scotland) of engineers, developers and planners (McKissock *et al*., 2003a).

4.2 LAND-TAKE FOR NEW DEVELOPMENTS

The land-take requirement for stormwater treatment facilities (BMPs) varies with the type of system selected (see Chapters 2 and 3). For example, end-of-pipe facilities such as ponds and wetlands occupy space that could arguably be otherwise allocated to additional development. Source control facilities however, offer the opportunity for stormwater management in dense developments, especially where roadside filter drains are utilized, as in parts of Scotland (McKissock *et al.,* 2001), and in Japan (Fujita, 1994b; 1997). Ultimately it is a matter for the urban authorities (city, water utility and/or environment agency, according to local jurisdictions) to require BMPs, with appropriate technical guidance on selection of drainage options according to type of development and local circumstances and conditions, as set out in Chapter 3. Once requirements are established then developers seek only a 'level playing field' for their businesses in allocating land for BMPs. That can, of course, be very difficult to achieve when local circumstances can lead to different land take requirements.

4.2.1 Densely developed urban areas

There is negligible additional land-take in the Japanese infiltration curbs and filter drains represented in Figure 4.3. The figure shows an example drainage system for a car park and associated buildings. Together with a detail vertical section of an

Text Box 4.1 Perceptions, problems and misunderstandings

Barriers to the introduction of innovation into working practice can be classified as follows:

- technical problems
- institutional traditions and human behaviours

For technical problems there are usually technical solutions – certainly in the field of stormwater management, where sufficient research has been done and practical experience gained (in several parts of the world, in a variety of hydrological conditions) to provide a sound basis for design and operational decision making. (That is not to say there is no need for further research and monitoring, as discussed in Section 4.8 below). The technical problems arise when developers and the associated planners, engineers, landscape architects and regulators, do not have sufficient awareness of the technical literature and/or first-hand experience of the technology in practice. It is not uncommon in such instances for the initial design concept to be flawed. Examples are the use of swales and filter drains as end-of-pipe treatment options for conventional highway drains and run-off from separately sewered housing developments. Both sorts of stormwater management facilities have been built and failed in Scotland, for example, prior to the production of a locally available design manual, and the accumulation of sufficient first-hand experience by local engineers and regulators etc. (D'Arcy and Roesner, 1999, and Jeffries *et al* 1999 and 2003).

Many aspects of apparently technical issues are a direct result of institutional or behavioural constraints – a reluctance to consider technical guidance from another country is a good example. Or a simple refusal to consider something different. Even where engineers and others are interested, they can be constrained by regulations or policies; institutional or statutory change is often necessary.

infiltration drainage system installed beneath the street to collect and infiltrate the surface flow (Fujita, 1994b). Important features are the bucket at the inlet to collect materials that might clog the infiltration network, and the sediment trap below it (Fujita, 1994a), see Figure 4.3a.

The use of permeable surfaces for car parks and pedestrian areas within towns can provide a neutral land-take option for development. As well as in Japan, examples of this type of water quality treatment BMP (that also attenuates flows – see Sections 4.6 and 4.7 below) are especially popular in the UK and France where their effectiveness has been well studied and demonstrated over several years (Pratt, 1999a; Macdonald and Jefferies, 2001; Abbott and Comino-Mateos, 2003; Raimbault, 1999). Underground storage using shallow systems with load-bearing capacity but substantial void space is another effective option where space for flow attenuation is minimal (Andoh *et al.,* 2000).

Text Box 4.2 Infiltration of stormwater: source control in Japanese cities. (from Fujita, 1994a; 1994b; 1997).

Infiltration facilities were introduced to Japan in the 1970s and implemented progressively during the 1980s. Soil conditions in Tokyo are favourable for infiltration. Initially the focus was on reducing run-off volume, but latterly positive benefits for groundwater recharge and dependent life in the cities are much appreciated. The first time that permeable asphalt pavement was introduced to Japan was to supply sufficient amount of water to trees lining the paved streets. Because its strength is lower than ordinary asphalt, the use of permeable pavement is restricted to sidewalks and other pedestrian areas where there is traffic volumes are low. All the footpaths in Tokyo have been changed to permeable asphalt since 1983, and the use of permeable asphalt continues to spread steadily. A variety of infiltration devices have been developed in Tokyo: (1994 figures)

Infiltration inlets	87,804 units
Infiltration trenches	626 km
Infiltration kerbs	123 km
Permeable pavement	4.439 km^2.

Permeable concrete block pavement is also used in Tokyo. Permeable asphalt and concrete pavement are both popular with Tokyo residents because of the enhanced safety (non-slip) and cleanliness (no puddles). Rainwater from roof areas is drained to soakaways at housing sites, and infiltration inlets are installed in the streets to infiltrate stormwater run-off from roads.

With 24,740 houses in an area of 11.32 km^2 (1994), the city of Koganei (15km west of the centre of Tokyo) is an example of success in reviving lost natural environments within a city by the widespread use of soakways. Many natural springs have reappeared and local residents have been providing household soakaways at their own expense. Between April 1983 and December 1994, up to 15,000 soakaways in 3,700 sites were provided by residents. This technology has also been implemented in other towns and cities in Japan, including Sapporo, Shiogama, Chiba, Yokohama, Nagoya and Amagasaki (Fujita, 1997).

4.2.2 Finding space for site and regional controls

Site controls can be more difficult – although it may be possible to accommodate swales and/or detention basins within the greenspace requirement of a new development, which is often 5–10 per cent of the development area. In Scotland that is encouraged in government guidance for developers and local authorities. Much of the greenspace, however, may already be allocated for other land uses in the minds of the developers and landscape architects; early discussions maximize

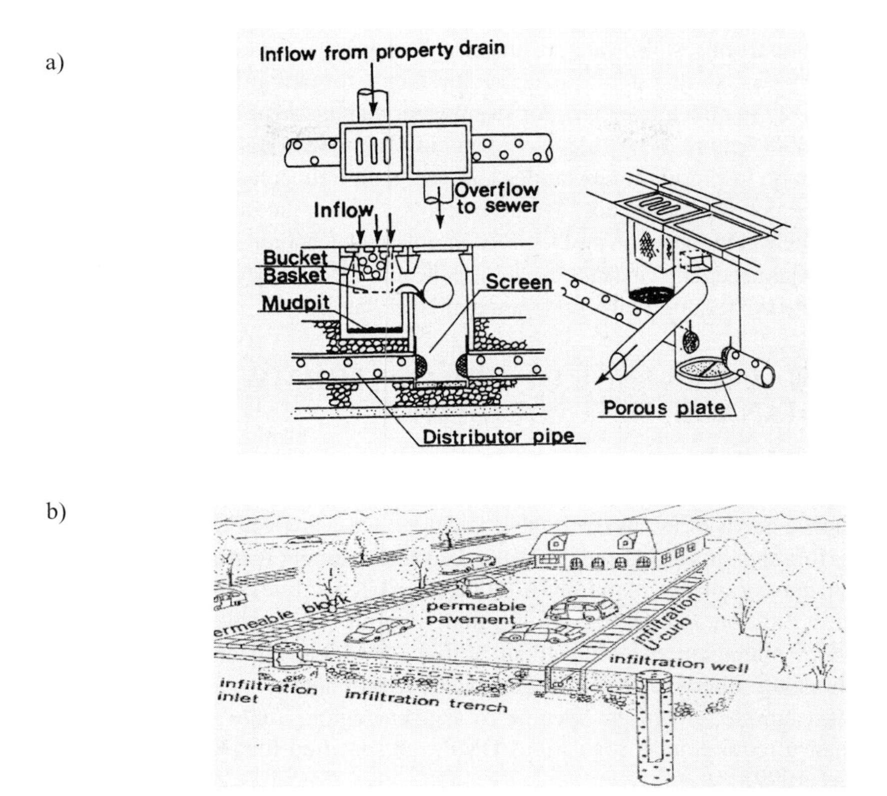

Figure 4.3 a) Infiltration of inlet for street drainage (Fujita, 1994a); b) Flow attenuation system in Tokyo, Japan: example lay-out, and vertical section of roadside infiltration drain (Fujita 1994b).

compromise opportunities. Many countries have developed innovative ways to fit more sustainable stormwater systems into urban areas. Schueler identified the merits of housing schemes involving clusters of dense housing units, separated by greenspace, as a means of reducing the impervious area per head of population, and also providing areas of greenspace in which stormwater management facilities (BMPs) can be located. At its most extreme, that would encourage high-rise flats; although unfashionable for housing (in the UK at least) the idea has a lot to offer for offices and other buildings.

For suburban areas, where properties have gardens at the rear that meet to form a common boundary, that resulting line can be the course of a collection system such as a filter drain or swale. This has been done in Sweden, USA and other countries. It obviously could be constrained by levels and other local circumstances, but should not be constrained by disinclination to consider the idea.

On a regional scale, stormwater treatment or management facilities can be planned as wetland parks, so that public space requirements for new urban developments can be utilized in allocating space for stormwater management facilities. That has been a notable feature of new developments in Malmö and Halmstad in Sweden, and in Orlando in Florida for example. The treatment train concept means that the more source control techniques that are employed within the urban development, the better the water quality in any regional facility (see Chapter 3 and Section 4.9). But that means accepting land take can be significant.. Table 4.1 gives an approximate idea of land-take requirements for a range of BMP facilities.

4.3 MAINTENANCE OF URBAN STORMWATER MANAGEMENT FACILITIES

Considerable experience has accrued in several parts of the world where BMP technology has been applied over the last 10–20 years. A detailed consideration of maintenance needs of urban BMP facilities in the USA is given in Schueler *et al.* (1992), Horner *et al.* (1994) and the Watershed Management Institute (1997). Less comprehensive guidance is in CIRIA (2000) (for UK facilities). Its importance is stressed in Larsson and Karppa (1999); underpinning the notable successes with stormwater management achieved in Malmö, Sweden. No drainage system, conventional or otherwise, has no maintenance requirements.

As an example, albeit site specific to a given region, Table 4.2 summarizes maintenance requirements for various BMPs, as specified for Denver, Colorado (Urbonas, 1999).

Maintenance requirements are driven by two primary considerations:

- maintaining functionality
- amenity.

The former requirement is primarily concerned with the periodic removal of the pollutants that the facility is designed to trap in order to prevent pollution of the aquatic environment: primarily sediment and litter. For vegetative systems there is also a variable requirement to remove excessive vegetative growth and systems alongside roads (swales and filter drains) require occasional minor restoration in places when vehicles damage their edges..

On some ponds it may be necessary to remove algae or weed, where local communities feel strongly that such natural characteristics are unacceptable in their (usually very formal) suburban landscapes. In Edmonton (Alberta, Canada), conventional ‘parks and gardens’ style landscaping and maintenance around the Beaumaris stormwater lake has created a tidy and attractive amenity, popular with cyclists, joggers and young families who come to feed waterfowl. But there’s not much natural vegetation in the landscaping specification and consequently a lot of

Table 4.1 Example landtake requirements for BMPs (based on an M5–60 minute rainfall depth of 15mm with relatively impermeable clay, or loam over clay soils)

BMP facility	*Land-take as percentage of catchment area*	*Comments*
Retention Pond a) 50% impervious cover in catchment area b) 25% impervious cover in catchment area	 3.5% 2.5%	This BMP facility, which can provide biological treatment has been assessed using an average depth of 1.25m.
Stormwater Wetland a) 50% impervious cover in catchment area b) 25% impervious cover in catchment area	 8.25% 5.75%	This BMP facility, which can provide biological treatment has been assessed using an average depth of 0.4m.
Extended Detention Basins a) 50% impervious cover in catchment area b) 25% impervious covering catchment area	 2.5% 1.75%	This BMP facility, which cannot provide biological treatment has been assessed using an average depth of 0.5m.
Permeable Pavers (including attenuation underground, either in stone fill, or alternative media reservoir)	No additional land required	Permeable pavers plus stormwater storage for car parks and pedestrianized urban areas can accept all incident run-off, plus some from associated buildings and roads.
Swales Source control, along roadside(s)	Up to 3m more than conventional drainage alongside road edge.	May be scope for utilizing road edge landscape zone in towns, shouldn't be an issue in rural areas, subject to topography and local conditions.

algal problems, requiring regular treatment with herbicides. That comprises a major part of the annual maintenance costs. In Calgary, by contrast, the city authorities are more in favour of natural-looking ponds and wetlands (and low maintenance costs) (D'Arcy, 1998a). But the local public has to be comfortable with the more natural or semi-natural look. Such acceptance has been well managed in Malmö and Halmstad in Sweden.

Additional maintenance needs arise occasionally according to the exposure of the facility to damaging influences. For example swale edges and gravel surfaces of filter drains are liable to disruption when heavy vehicles occasionally stray off the road into them. These instances can be avoided by providing kerbs (open-jointed) or (preferably) rumble strips (high noise plastic bumps along the road edge frequently used along motorway margins in UK).

Table 4.2. Example maintenance requirements for urban BMPs (Urbonas, 1999).

Maintenance requirements	*Permeable surface (PS)*	*Grass Buffers (GB)*	*Grass swales (GS)*	*Porous landscape detention*	*External detention basins (EBD)*	*Sand filter basins (SFB)*	*Retention ponds (RP)*	*Construct wetland channel*	*Maintenance objectives and considerations*	*Frequency*
Inspections	✓	✓	✓	✓	✓	✓	✓	✓	Prevent deterioration. sediment accumulation inlet/outlet blockages poor infiltration hydraulic performance; cover and condition of grass vegetation; structural integrity/damage; erosion	Annual or bi-annual. Inspect system during or following major storms to monitor performance
Litter and debris removal	✓	✓	✓	✓	✓	✓	✓	✓	Source control: Removal: aesthetics; prevents down stream flushing & effects such as clogging/blockage	Routine. Preceding and following storm seasons frequency depends on aesthetic requirements
Structural repairs	✓	✓	✓	✓	✓	✓	✓	✓	Identify and remediate: areas affected by erosion (GB; GS; EDB; RP; CWC) waterlogged areas. Inspect system during or following major storms to monitor performance	Non-routine. Dictated by inspections
Lawn care, mowing, vegetation care		✓	✓	✓	✓		✓	✓	Maintain dense grass cover (lengths: irrigated grass 50–100mm; non-irrigated grass 150–200mm). Collect and dispose of cuttings offsite or use mulching mower	Routine. As dictated by inspections
Sediment removal			✓	✓	✓		✓	✓	Remove accumulated sediment near inlets and in channels or bottoms to maintain flow capacity. Frequency may need to be increased where construction is ongoing in the catchment	Routine. GS: 3–10% of total length/year. PD; EBD: inlets every 1–2 years. PLD; EDB; RP; CWC: bottom/bed every 10–20 years

Maintenance requirements	*Permeable surface (PS)*	*Grass Buffers (GB)*	*Grass swales (GS)*	*Porous landscape detention*	*External detention basins (EBD)*	*Sand filter basins (SFB)*	*Retention ponds (RP)*	*Construct wetland channel*	*Maintenance objectives and considerations*	*Frequency*
Erosion control		✓	✓		✓		✓		Repair and revegetate eroded areas in detention basins & swale channels, repair damaged inlet/outlet energy dissipaters	Non-routine. Dictated by inspections
Nuisance control					✓		✓	✓	Address problems such as odour and overgrowth associated with stagnant or deoxygenated water, using appropriate measures.	Non-routine. Dictated by public complaints and inspections
Irrigation		✓	✓		✓				To maintain required min. soil moisture for dense grass or vegetation growth Only likely to be necessary during establishment of vegetation or during very dry periods	Infrequent. As dictated by inspections
Scarify surface		✓	✓						To encourage growth of grasses following desiltation (swales and filter strips); to promote drainage in sand filter infiltration basins-rake filter surface (top 100–150mm)	Annually (sand filter infiltration basins); as necessary (swales and filter drains)
Turf replacement grass reseeding and mulching		✓	✓						Maintain healthy dense grass cover. Turf alterations or replacement should only be necessary where water flows are restricted (e.g. turf height above pavement level) or where turfs are degraded (e.g. by vehicles, erosion, siltation effects of pollutants)	Non-routine. As dictated by Inspections

For ponds and wetlands, estimates of the de-sludging frequency vary: a worst-case scenario adopted for guidance in the UK suggests that silt needs to be removed from ponds (in order to sustain effective pollutant removal and flow attenuation capacity) after some seven years of normal operation; (CIRIA, 2000). Other published sources suggest every 15–50 years (Fairhurst & Partners, 2002), and also see Table 4.2.

It is widely recognized that the peak sediment load on a stormwater pond is during the initial phases of development, when soil is usually exposed to wind and rain erosion. Suspended solids loads as high as several thousand mg/l have often been recorded in construction site run-off. Once a site is fully developed, solids loads in run-off are typically much lower, of the order of tens or hundreds of mg/l. It is therefore especially important that drainage facilities installed to protect the aquatic environment during the construction phase are reinstated once development is complete, in order to maximize the life of the new facility.

For permeable pavements, Fujita (1994a) reported that clogging of permeable asphalt normally occurs to a depth of about 3mm, and a cleaning machine has been developed and is widely used effectively. High pressure water strikes the surface of the pavement to keep it clean and the practice restores the capacity of the clogged pavement to that of new pavement (see Figure 4.4).

In the UK, experience at motorway service stations where SUDS (urban BMPs designed for flood and quality control purposes) have been installed, is that maintenance can be less expensive than for conventional drainage systems (Bray, 2001b).

Figure 4.4 Cleaning machine for permeable pavement (from Fujita, 1994a).

Text Box 4.3 A landscaping specification for urban BMPs

Inappropriate planting around a stormwater management facility can jeopardize the achievement of its water quality and flow control functions. An example is the blockage of drainage pathways between mono-block pavers on a permeable surface car park, by soil washed off surrounding embankments. Where the landscaping specification has required non-native shrubs, with bare soil between each plant, significant loadings of topsoil wash off onto the adjacent car park surface to cause local loss of porosity. This has also been observed for open pore tarmac car parking bays. The problem is not resolved by mulching with bark, wood chip or other surface covers; only a well-established grass sward will stabilize soil on slopes sufficiently to protect the functionality of the permeable surface.

The problems are not limited to permeable surfaces, but can also be an issue for detention basins, swales and retention ponds, where the quantities of topsoil (and associated nutrients and pesticides) washed into the pond from the immediate surrounding landscaped area, may be comparable to the influent loadings of suspended matter that the facility is designed to remove. A native grass and wildflower seed mix will stabilize banks and surrounding landscaped areas, need less maintenance than conventional land cover, and not require applications of pesticides to maintain the vegetation. Such surrounds are also vastly more beneficial for wildlife than conventional, expensively maintained flower beds and exotic shrubs.

Figure 4.5 shows how inappropriate landscaping can result in excessive contamination of run-off by soil, together with some examples of good practice (for example the level of the soil in landscape features within or alongside permeable car parks must be kept safely below the height of the surrounding kerb (and a flat kerb is not sensible).

Where shrubs are required for landscaping purposes or as part of a barrier to access for children, there is always a choice of native species that will thrive within a native grass and wildflower soil cover. (Contributed by landscape architect Bob Bray, of Robert Bray Associates, Stroud UK)

Soil level below kerb height.

Poor practice: no kerb to retain soil.

Gravel drain added to intercept run-off from landscaped area that had eliminated the porosity in the permeable surface with soil wash off.

Block pavers being uplifted and soil removed to re-establish porosity once relaid.

Formal landscaping creates erosion risks:with possibly as much contamination arising from exposed soil around the pond as from the drainage area it serves.

Figure 4.5 Landscaping issues for effective performance of BMPs.

4.4 ADOPTION AND PUBLIC OWNERSHIP ISSUES

Many excellent BMP facilities are privately owned or managed, serving industrial or commercial premises or other developments not on the public drainage network. Such developments are not constrained by drainage specifications of local authorities (for example requiring kerbs and gully pots for roads) or policy limitations of water utilities (e.g. a decision to not adopt any ponds). Such independence allows innovation and the opportunity to realize potential cost savings (see, for example, Figure 4.7 which is based almost solely on uptake by private developments).

Nevertheless, most of the good examples of BMPs in this chapter are in public ownership, or at least part of a public drainage network. Examples are the BMPs in Malmö and Halmstad in Sweden, in Tokyo (Fujita, 1994a) in Japan, in Orlando, Florida, in Austin, Texas (Richards, 1999) and other USA towns and cities. A single municipal authority has obvious potential advantages: drainage, roads, planning, parks and gardens functions all under the same metaphorical roof. In practice however, the same consensus building process is still needed as for developing acceptable schemes in situations where several public authorities and agencies are involved (as, for example, in the UK, see Conlin, 2000,and D'Arcy and Harley, 2002).

The critical issue is design standards for developers to follow, if they hope to have their infrastructure adopted. In the USA this is achieved via BMP manuals in the various states or cities (e.g. California, Maryland and Denver, Colorado). In the UK, since 2000 designs are to be in accordance with manuals published on behalf of stakeholder groups (eg CIRIA, 2000). In January 2003, in Scotland statutory responsibility for adoption and maintenance of public stormwater systems was given unequivocally to the water utility, Scottish Water, putting the term 'SUDS' (the same measures as BMPs, but with aspirations for integration with flooding concerns and optimization of amenity opportunities, see Section 4.7) into law.

A major benefit of clear responsibilities for key infrastructure features is of course that the environmental regulator has a body with whom to negotiate retrofit programmes to address historical environmental degradation of urban watercourses (see Section 4.6). In Scotland for example, some 500km downgraded due to urban run-off in Scotland, including about 150km associated with industrial estates (SEPA, 1999).

There remains a need for a spirit of partnership in providing BMPs in the urban landscape – especially if the amenity and habitat aspirations of the SUDS concept are to be realized. For example BMPs are facilities to deal with run-off from small frequent storms 'water quality storms'; but the wider concerns of flooding associated with infrequent major storm events also need to be considered in planning the urban environment. The creative use of public greenspace to integrate these different but compatible objectives, in a landscape that provides some amenity value as well, is the challenge facing developers, planners, water utilities or local municipal authorities, and environmental regulators.

Text Box 4.4 SUDS For Scotland

The Water Environment and Water Services Act received Royal Assent on 3 March 2003 and amended the Sewerage (Scotland) Act 1968 to include a definition of SUDS as follows:

Sustainable urban drainage systems *(in Scotland)*
"SUD system" means a sustainable urban drainage system;
"sustainable urban drainage system" means a drainage system which—
(a) facilitates attenuation, settlement or treatment of surface water from 2 or more premises (whether or not together with road water), and
(b) includes one or more of the following: inlet structures, outlet structures, swales, constructed wetlands, ponds, filter trenches, attenuation tanks and detention basins (together with any associated pipes and equipment);.

The amendments also provided clarity for the role of Scottish Water:
"private SUD system" means any SUD system which is not a public SUD system;
"public SUD system" means any SUD system which is vested in Scottish Water.

This gives public SUDS the same validity as traditional sewers. Those measures clarify the scope of the water utility to adopt and maintain SUDS in Scotland, ultimately recovering costs from customers. SUDS has become established technology, and the standards to which facilities must be built will be set out in a document supporting detailed government regulations: *SUDS for Scotland*.

4.4.1 Public safety

Included in any public document giving design standards, should be guidance on safety issues associated with water. In the UK reference is made to the ROSPA (1999) publication: *Safety at Inland Water Sites*. That report highlights rivers (30%) and coastal waters (24%) as the major categories of drowning incidents. Garden ponds (2%), canals (8%) and docks and harbours (5%) are the categories most similar to urban stormwater ponds and the lower numbers reflect the lesser risks with still water bodies. The key message is to teach water safety to children from an early age; BMPs are a part, but only a small part of the risks the public must manage, in the course of enjoying the environment. Detailed safety considerations are set out in several BMP / SUDS publications (e.g. CIRIA, 2000) and typically follow Schueler (1987) who states that wet ponds can be designed to minimize the risk of accidental drowning by keeping them relatively shallow, installing an underwater safety bench, avoiding any sharp drop-offs from shores, keeping side slopes gentle, and fencing off large diameter outfalls.

Some local authorities require fences around ponds; a 'toddler fence' is a good idea – able to deter very young children but not high enough to prevent a parent or grandparent getting over it if necessary to rescue an unsupervised child in difficulty. The toddler fence concept recognizes that older children will get over any fence if they want to, unlike many adults. Useful accounts of safety issues are also given in Ferguson (1998) and in Campbell and Ogden (1999).

4.5 COSTS ASSOCIATED WITH URBAN BMPS

The costs of BMP drainage facilities as compared with conventional systems is often difficult to separate from other factors, such as the development proposals and vision of the developer, the regulatory regime influencing developers, the charging schemes imposed by local councils in relation to stormwater collection and disposal, and the criteria for adoption of drainage networks by the public authorities or agencies.. Comparisons between countries are difficult because of the differing systems of responsibility for managing stormwater in various countries. In Sweden, it has been said (Stahre, 1996) that BMPs are almost invariably cheaper than conventional systems; that is possible because local problems with the drainage system are costed for solution by the people directly served by the problem part of the network. It is therefore in the interests of the public to consider the most cost-effective solution (Peter Stahre, in the video *Nature's Way*, 1996; see Text Box 4.5, and Figure 4.6). By contrast in the UK, the cost would be spread over the entire contributing population, not those directly affected and the cost-effectiveness and acceptability of the solution is lost. Furthermore, BMPs (SUDS) have often been an extra cost because local authorities have only accepted them in development proposals if they are an add-on (additional) feature alongside (or literally at the end of) a thoroughly conventional system. For example a housing development might be conventionally drained, complete with surface water sewers, road gullies, kerbs and mown lawns alongside the roads and in roundabouts. Then a BMP facility has to be added and a house plot or two has to be surrendered for provision of a pond or other end-of pipe drainage facility.

Where the BMPs can be considered at the outset, as alternatives to conventional drainage systems, real cost savings can often be achieved, especially for private developments where there are no constraints associated with local authority requirements for standard services (e.g. Bray 2003 and 2001b).

Text Box 4.5 Realizing cost savings – a Swedish case study

In Malmö, in the extreme south of Sweden, a part of the city served by a combined sewer suffered from flooding in the basements of houses in several streets during wet weather. The city estimated the cost of providing a new relief storm sewer (that would have required a higher frequency of spills of storm sewage into the local watercourse). That cost was put to householders and amounted to several thousand kroner per household. An alternative solution was adopted, whereby each household was offered a payment of a few hundred kroner to divert roof run-off out of the public sewer and onto the garden. Rain water collection and storage systems were sometimes utilized too. Overflow water (or direct run-off from the roof) was led onto lawns by a couple of metres of paved spillway, supplied by the city (see Figure 4.6). Faced with such a choice of costs to solve the flooding problem in the houses, householders chose the disconnection system, saving the city the capital outlay required for a new sewer. Interestingly, the soils are not very porous. That lack of porosity probably explains why the household basements do not flood due to the ingress of water from the roof via the gardens. The peak flow of run-off is believed to be attenuated by the garden system, even if it eventually finds its way back into the sewer. Householders are prepared to accept wet gardens in wet weather, because it solved an existing more severe problem and was a zero-cost solution.

Peter Stahre, referring to experience in the city of Malmö, states that BMPs are almost invariably less expensive than conventional alternative technologies; differences in favour of BMPs ranging from 15 per cent savings to considerably more. The BMP sites in Malmö are described in Larsson and Karppa (1999) who conclude that BMPs are a cost-effective means of achieving both quantity and quality control of urban stormwater run-off.

In the IAWQ (now IWA Publishing) video *Nature's Way* (IAWQ, 1996 and now reissued on DVD with this publication) Torsten Rosenqvist, for Halmstad and Tom Schueler for Maryland both report that houses overlooking stormwater wetlands attract higher prices than similar nearby housing without such benefits. That has also been the experience of some UK developers (e.g. Jock Maxwell in the UK urban drainage video *Designs that Hold Water*). Not all developers report such findings however and equivalent anecdotes attest to the failures of some BMPs to appear attractive to house-buyers (or possibly other factors; independent surveys have shown a positive interest (Apostalaki *et al.*, 2001, and references quoted in Schueler, 1987).

A major part of the cost of creating a BMP facility can be planting (e.g. for ponds and associated landscaped areas) – at one UK site that exceeded the costs of

Figure 4.6 Diversion of roof water to lawns in Malmö, Sweden.

the excavation and preparation of the pond site. It is good practice to allow natural colonization wherever possible, restricting planting to barrier species (such as *Phragmites*: often required as part of planning authority approval for developers). Best practice can therefore be less expensive, but native vegetation in the ponds and the adjoining areas may need to be 'sold' to the local authority, general public and the developer. Detailed planting advice is given in SEPA, 2000a, and Campbell and Ogden, 1999. There will undoubtedly still be circumstances where it is an expense.

For the UK, an indication of how cost effective some BMP options can be when truly an alternative to conventional techniques has been the number of permeable pavement systems sold by a manufacturer of specialist monoblocks with patented permeable surfacing systems including sub-base specification for stormwater attenuation, (for Hart 2003, pers. com. see Figure 2.5). In the first eight years after introducing the product to the hitherto unexploited UK market in 1997, sales grew steadily; see Figure 4.7. Eventually rival products also appeared on the market in

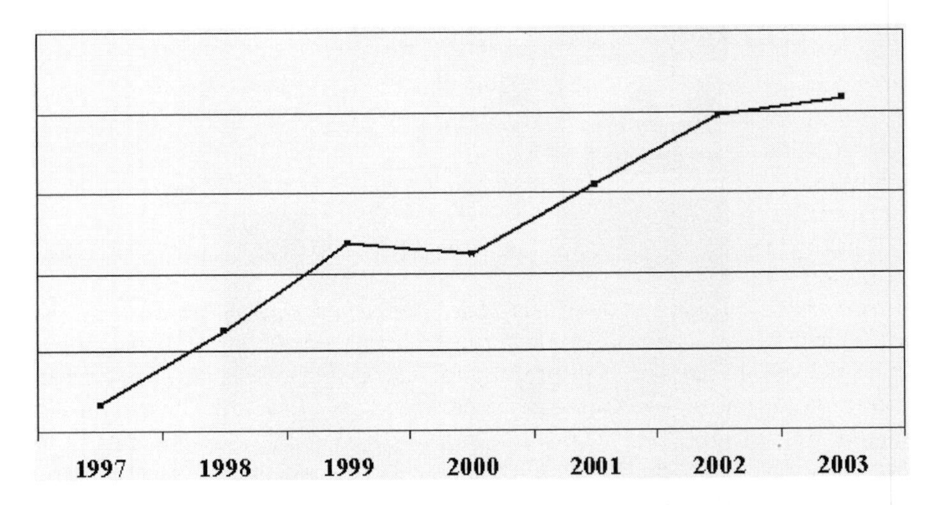

Figure 4.7 Sales of Formpave Aquablocks in UK, since introduction to the UK market as part of a new permeable surfacing system for car parks and driveways. The *Y* axis values are equivalent to over 500 car parks by 2003. Except in south-east Scotland, for the initial years there was no strong regulatory drive for this technology; it is therefore a measure of cost-effectiveness that it has been established in the UK (from Holt, 2003).

the UK. The important point for consideration of costs of BMPs is that the most sales were in parts of the UK where the regulator was not driving the application of BMP technology for stormwater (from 1995/6 to 2003, a regulatory drive for us of this technology was largely restricted to Scotland, initially in the south-east). Uptake in England and elsewhere was on the basis that the technology has been cost-effective to meet the needs of the client developers. A 10–15 per cent net saving on the cost of providing a drained car park is claimed as typical based on absence of gully pots and grids, surface water sewer network, or oil interceptors (Hart, 2003, pers. com.).

The bottom line on costs, however, is simply that the prevention of environmental damage should be a key feature of development proposals, and sustainable drainage solutions should be a requirement.

The cost of retrofitting urban BMP stormwater facilities has been considered by Lindsey and Doll (1998), particularly looking at the role of stormwater utilities. They consider several revenue options including a comparison of property taxes and stormwater user charges, arguing that user charges are preferable to taxes because they provide the user with an incentive to reduce the amount of impervious area on their property.

4.6 RETRO-FITTING URBAN BMPS

The application of best management practice technology in existing urban environments is driven by any or each of the following:

- Brownfield site redevelopment
- Reducing the frequency and impact of combined sewer overflows on receiving waters
- Restoration of receiving water quality in streams adversely impacted by urban run-off

Brownfield sites are areas of land that have been developed already and are available for redevelopment. Such land may be contaminated, especially if a former industrial site, or may simply be property for redevelopment within existing urban areas. Many brownfield sites are in the older parts of towns and cities, typically drained by a combined sewer system. BMPs can be used to reduce pollution from combined sewers, by attenuating run-off and reducing peak flows into the sewer. Finally, many newer urban conurbations are served by separate sewer networks and they result in pollution by the transfer of pollutants to watercourses (especially associated with road traffic, but also other urban sources of contamination, see Chapter 1, and for pollutant values Choe *et al.,* 2002 and Mitchell, 2001). BMPs offer a prospect of restoration for some urban streams, in conjunction with other measures, (see treatment train, Section 4.9 below).

In California, an extensive programme of BMP retrofits was initiated in 1997, in Los Angeles and San Diego, with the objective of determining the cost-effectiveness and water quality benefits of structural BMPs. Construction began in September 1998 and was substantially complete in March 1999. Thirty-nine BMPs were constructed, 26 in LA (at 21 sites) and 13 in San Diego (at 12 sites). Total construction costs are estimated at $9 million. The sites served included freeways, interchanges, park and rides, and maintenance stations. (Nine types of BMPs have been installed and a monitoring programme was planned from March 1999–2001. Full details are available on the Caltrans website (see Caltrans statewide water quality index): http://www.dot.ca.gov/hq/env/stormwater/index.htm

Whilst not strictly *retro-fitting* BMPs, in the sense of adding facilities to serve existing developments, it is worth noting that a slower rate of BMP retro-fix is occurring continuously, once a local area establishes policies to require BMPs for new development and for redevelopment. Approximately half of the 767 sites in the Scottish SUDS database, for example, were brownfield sites; redevelopment of urban or former commercial properties (examples below). In the absence of multi-million dollar schemes such as the Caltrans project above, or rather the limitations of budgets and the need to address priority areas first, that slower rate of BMP provision (a function of redevelopment rate) will be the main implementation process in many urban areas.

4.6.1 Brownfield redevelopment

Applying best management practice technology on redevelopment sites within existing urban environments, especially high density housing or industrial units, can be very difficult, but there are BMPs available, as described in Section 4.2 earlier. The principal constraint is available space; for example broad swales and extensive ponds are not easily fitted into brownfield redevelopments in dense urban areas. Other BMP options have neutral or minimal land-take, however, and have been successfully integrated into existing developments. Macdonald and Jefferies 2001, describe the performance of a permeable paved car park on a redeveloped site in the city of Edinburgh (see Figure 4.8). The technology is now recognized in the UK as a cost-effective drainage solution. A major motor car retailer in the UK is increasingly using such pavers on car parks at sales outlets, for the convenience of customers and the maintenance of clean car interiors by eliminating dirty-water puddles, as much as for pollutant and storm flow attenuation. Other UK retail outlets are also adopting permeable pavers for surface water drainage at new or redeveloped urban areas (e.g. major supermarket chains). In the UK over 500 contracts have been placed with the principal supplier of permeable block systems in the eight years since the idea was introduced from France (via the Hydro International publication *The Natural Way*). A significant proportion of that number is for urban redevelopment (Hart, pers. com. 2003).

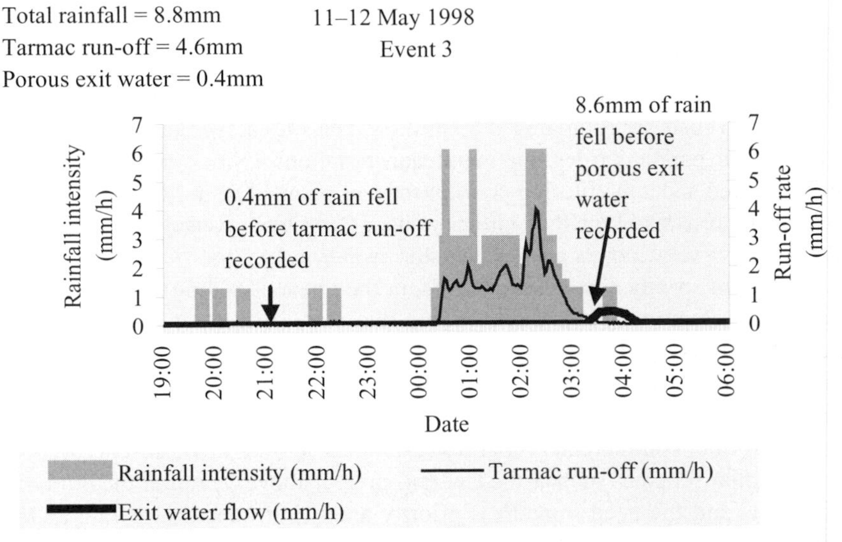

Figure 4.8 Performance of permeable pavement compared with conventional tarmacadam car park. (Macdonald and Jefferies, 2001).

4.6.2 Reducing the frequency and impact of combined sewer overflows on receiving waters

For flow attenuation only, various proprietary devices have been developed to constrain peak run-off from conventional impervious surfaces. These can have significant pollution prevention benefits in existing urban areas served by combined sewers and associated storm sewage overflows to watercourse. More traditional measures such as tank sewers and off-line storage tanks can also be employed when redeveloping existing urban areas, although the availability of existing sewer connections in such redevelopments can favour simple reconnection of drainage at minimal cost, but no strategic benefit. The permeable surfaces noted in the preceding section (4.6.1) can also be very effective in reducing peak flows into a combined sewer network (Macdonald and Jefferies, 2001). Fujita (1997) reports that a ten-year strategy of promoting such technology has been effective in Japan at reducing CSO (combined sewer overflow) problems.

In the high-density urban parts of Tokyo City, it has been estimated that a reduction in stormwater flows by adoption of source control techniques could reduce the number of CSO spills for a given rainfall year to 19 per cent only a fifth (approximately) of the number of sewer overflow discharges during the study year from the same catchment with conventional drainage (Fujita, 1994a). Similar investigations have been undertaken in the UK; (Stovin and Swan, 2003). McKissock *et al* (2003b) identified the location and relative importance of areas of imperviousness in the Scottish town of Dunfermline, together with the distribution and extent of green space and pedestrianized streets. The likely costs and problems of SUDS retro-fits where compared with the costs of conventional storm sewage management there with the potential for retro-fitting SUDS (urban BMP facilities). Conventional measures in Dunfermline involved massive expenditure and land-take in building very large storm tanks, complete with storm pumps to return stored sewage to sewer, and automatic self-cleaning screens to treat sewer overflows of effluent prior to discharge to watercourse. With hindsight it would probably have been cost-effective to retro-fit BMPs if agreement could have been secured for the necessary source and site controls for storm water management in the combined sewer catchments.

4.6.3 Restoration of receiving water quality in streams adversely impacted by urban run-off

In several parts of the world projects have sought to restore urban watercourses by combinations of measures, involving BMPs for urban drainage and river restoration techniques. Japanese research work has contributed much to the characterization and quantification of diffuse pollution from urban areas, (e.g. Haiping and Yamada, 1996; Uchimura *et al.*,1997); that knowledge has informed the process of developing BMPs to resolve urban problems. The Japanese experience with hard engineering

BMPs has already been referred to in Section 4.2 and Section 4.6.2 (Fujita, 1994a and 1997). But as well as contributing to reduced CSO spills, the retro-fitting of BMPs in Japan has also evidently contributed to water quality improvements, by retaining run-off pollutants such as sediments, toxic metals, oil and PAHs. Yamada *et al.* (2002) considered urban sources of nutrients (in relation to studies at Lake Biwa, in Japan) and investigated the efficacy of soils (the key interface with pollutants from road traffic in many BMPs) at adsorbing total nitrogen, total phosphorus, and COD showing that direct run-off into roadside soils led to reductions in pollutant loads of 80 per cent for COD, 50 per cent for T-N, and 40 per cent for T-P. Addressing roads as diffuse sources is, of course, only a part of the effort to restore Lake Biwa (Fujii *et al.,* 2002; Ichiki and Yamada, 1999).

Despite apparent difficulties, retrospective drainage solutions have also been attempted with ponds and other soft-engineering options, sometimes in combinations with a variety of measures. Several watercourses are involved in the San Diego part of the Caltrans BMPs Retrofit Pilot Studies project, referred to at the start of this chapter (see Table 4.3). Other factors affect water quality there, however, and the aims of that programme are more focused on the performance of the BMPs as individual units. Including monitoring, the total cost of that programme was $30 million.

Table 4.3 Example BMP retro-fit facilities constructed in San Diego, California (1999; www.dot.ca.gov/hq/env/index.htm).

BMP type	*Location*	*Cost*	*Drainage area*	*Receiving water*
Exterior detention basin	Manchester (east)	$360,000	4.8 acres	San Elijo Lagoon
Exterior detention basin	SR 56	$166,000	5.3 acres	Los Penasquitos Lagoon
Exterior detention basin	SR 78	$855,000	13.4 acres	Escondido Creek
Wet basin	La Costa (east)	$694,000	4.2 acres	Bataquitos Lagoon
Infiltration basin	La Costa (west)	$241,000	3.2 acres	Bataquitos Lagoon
Media filter	Kearny Mesa	$340,000	1.5 acres	Tecolote Creek
Media filter	Escondido	$451,000	0.8 acres	Escondido Creek
Media filter	La Costa P& R	$242,000	2.8 acres	Bataquitos Lagoon
Media filter	SR 78 P & R	$231,000	0.8 acres	Buena Vista Lagoon
Bio swale	Melrose Dr.	$156,000	2.4 acres	Buena Vista Creek
Bio swale	Palomar Airport Rd.	$142,000	2.3 acres	Encinas Creek
Bio strip	Carlsbad (west)	$196,000	0.7 acres	Encinas Creek
Infiltration trench/strip with bio strip	Carlsbad (E)		1.7 acres	Encinas Creek

Florida has a long history of stormwater management to minimize pollution risks, and has run a programme of retrofit measures (Bateman *et al.*, 1998). Florida also features in the reviews of stormwater retro-fitting in England (1999) and England *et al.* (2000). Bell and Champagne (1998) have reported on Alexandria's urban retrofit programme, that involved seven retro-fit projects in 1998.

In many ways the run-off from industrial estates represents a worst case category of diffuse pollution (D'Arcy and Bayes, 1994 and D'Arcy *et al.,* 2000b); a priority for BMP retro-fits of that type, supported by treatment train of measures (see Section 4.8 below). Contaminated surface water drainage from an extensive industrial estate on the edge of the town of Livingston in Scotland, was responsible for the downgrading of 3.3 km of watercourse in the catchment of the River Almond (SEPA, 1999). An experimental pond and wetland water treatment facility has been created there and is reducing quantities of some of the pollutants in the discharge. It has resulted in an improvement in the quality of the burn from class D to C (bad to poor quality, SEPA river classification scheme). The wetland (a pond and marsh/reed bed) was designed prior to the publication of the UK design manuals (.e.g. CIRIA, 2000). For it to be fully effective, the size of the facility needs to be increased, and additional features provided within the existing landscaped areas upstream in the industrial estate (for example lengths of swales along some roadsides, and local detention basins to reduce hydraulic load and manage irregular oil and other pollution incidents). Table 4.4 presents data for the quality of the contaminated surface water discharge for Houston Industrial Estate, and the reduction in concentrations of pollutants achieved by the treatment wetland.

In Sweden a retro-fit programme is underway for Flemingsbergsviken, south of Stockholm (Larm, 1999). Stormwater treatment facilities (ponds and wetlands) have been constructed to treat polluted run-off from the existing urban catchment, to try and help reduce nutrient loads to Lake Orlangen. Other BMP major retrofit projects have been completed (for example in Halmstad in Sweden) or are being planned, for example, the Rouge River, the Metro Atlanta Watersheds Initiative, and the Tualatin River in the USA (Murrray and Cave, 1999; Richards, 1999 and Jackson, 1999). Englesby Brook (Vermont) is described in Section 4.8 – monitoring effectiveness.

Table 4.4 Pollutant reduction achieved by the stormwater wetland serving Houston Industrial Estate near Livingston, Scotland (Henderson, 2001).

- Average results show SS & BOD/COD reduced by ~50%
- 46 samples taken over last 4 years.
- Inlet SS = 27mg/l; Outlet SS = 14.7mg/l
- Inlet BOD = 10.5mg/l; Outlet BOD = 4.1mg/l
- Inlet COD = 66mg/l; Outlet COD = 37mg/l

4.7 INTEGRATING REQUIREMENTS: FLOOD CONTROL, WATER QUALITY AND AMENITY

A holistic approach to stormwater management has been developed in Malmö and Halmstad in Sweden. In Malmö, the city engineers and planners refer to old military maps of the area prepared in 1815, to identify former wetlands. When new development proposals are being considered, opportunities are taken to reinstate natural drainage pathways and provide stormwater ponds in appropriate locations (Stahre in IAWQ, 1996). The Tofteness development on the upstream edge of the older, central core of the city of Malmö is an example, whereby a new wetland park has been created adjacent to housing areas and treating run-off from industrial, commercial and highway areas too. Water quality treatment (small, frequent storms) is provided by ponds within the wetland that are linked by vegetated channels. The entire wetland park can flood in major (infrequent, large) rainy storms. The wetland is popular with local residents and visitors.

A similar approach has also been taken 150km to the north at Halmstad, where a programme of creating wetlands was initiated for the town and surrounding farmlands with the primary aim of reducing nitrogen loads to the Skagerak/Kattegat (Halmstad Project, see Fleischer *et al.,* 1994). The Chief Technical Officer for the town is a landscape architect, Tosten Rosenquist, and he has encouraged sensitive landscaping of stormwater management features such as wetland parks and lakes, and even using roundabout 'dead-space' for detention.

Roesner *et al.* (1999) describe a similar holistic approach to stormwater management for a major development in Florida, and that expertise was utilized for a demonstration project in Dunfermline, Scotland (D'Arcy and Roesner, 1999), leading to the development of the 'sustainable drainage triangle concept' (D'Arcy, 1998b) whereby it is proposed that a stormwater management facility, irrespective of the primary driver for its creation, should be designed so as to maximize the benefits for all three aspects of the aquatic environment, water quality (pollution prevention and control), water quantity (flood control and maintenance of low flows), and amenity (public acceptability and informal recreational use, and biodiversity).

An example of a BMP facility that maximizes benefits for all these aspects is the Green Park Ponds in Orlando, Florida; built to protect raw potable supplies from pollutants by urban run-off, a very pleasant and popular wetland park has resulted (see Figure 4.9). Similar successes have been achieved in Malmö and Halmstad in Sweden.

4.7.1 Sustainable urban drainage systems (SUDS)

The urban BMPs concept, as explained in Chapter 2, is a rationale to address diffuse pollution. Strictly, urban BMP engineering facilities are therefore pollution control devices and do not have a flood control function, nor necessarily any amenity value.

Figure 4.9 Green Park Ponds, Orlando, Florida.

Land values and the demands of the public for recreation, as well as the need to maximize value for money, have driven best practice for stormwater towards the more holistic approach illustrated in Section 4.5 above. In the UK this has led to the concept of SUDS – sustainable urban drainage systems; stormwater management facilities that aspire to the aims of the sustainable drainage triangle (see Section 3.10). The concept follows on from two earlier innovative ideas for stormwater management in the UK.

Source control

For hydrologists in the UK this ideas refers to the disposal of incident rainfall as close to where it falls as possible. That can either be by infiltration, or attenuation by wetting vegetation and soil in a swale, for example, prior to surface flow off site. The approach was initially advocated by flood control engineers in the south of England and was originally seen as an opportunity to encourage recharging of aquifers in the south of England where water resources are fully stretched to meet water supply demands. It has also been investigated and pioneered by research at The School of the Built Environment, Coventry University, which has led to the extensive interest in hard permeable surfaces and soakaways/infiltration in the UK (Pratt, 1999a; 1999b), and the annual meetings of the Standing Conference on Stormwater Source Control organized by Chris Pratt from the university.

By eliminating (or greatly reducing) cross surface scour, such source control systems also have obvious benefits for the control of diffuse pollution. Research at Coventry University has indicated that minor diffuse source oil contamination of the permeable block pavers, for example, will break down by bacterial action within the system (Pratt *et al.*, 1999). Major spills will pass through the system, however, unless the system is modified and additional provision made for larger spills such as the volume of a diesel fuel tank for a car or van (Newman *et al.*, 2003).

BMPs

The best management practice concept was introduced to the UK by the IAWQ video *Nature's Way*, and includes the source control approach, but also provides for end-of-pipe options such as stormwater ponds and wetlands and, importantly, has the requirement that the systems are tried and tested technology – best practice.

SUDS

This third innovative concept means drainage facilities that aspire to minimal adverse impacts on the environment, by managing water quality and quantity issues in an integrated functional way. They involve facilities that can be publicly adopted and maintained as acceptable feature of the urban environment.

It is hoped that by invoking sustainability considerations, some of the barriers to use of best management practice technology can be overcome in local authorities and water utilities. The concept is outlined in Conlin (2000) and CIRIA (2000). It is also being extended to consider rainwater use (MacKissock pers. com.).

There is considerable scope for innovation in the development of the SUDS idea. The application of SUDS technology to address pollution for spills of storm sewage from combined sewer overflows (CSOs), e.g. Stovin and Swan (2003), has already been referred to in Section 4.5. Another example opportunity is in the greening of urban areas (as already demonstrated for the more innovative provision of BMPs in Florida or Sweden for example).

The social dimension to soft-engineering BMPs has already been referred to (e.g. Malmö, Halmstad and Florida in Section 4.5). But there is also considerable scope for amenity benefits arising from the use of hard engineering BMPs in the urban environment. For example, use of permeable pavers for pedestrianized city/town centres, as well as achieving the normal water quality aims of BMP technology could also attenuate peak flows – and would provide dry (no puddles, no splashes) surfaces for the public use of the area. Permeable block surfaces may also facilitate gaseous exchange for soil and roots of urban trees and shrubs, as well as allowing rain to reach the soil a positive consideration for use of such technology in some parts of Tokyo (Fujita, 1994a).

The passive nature of treatment processes, (and relatively low cost establishment opportunities for BMP facilities) are also more sustainable than many conventional technologies. But cost savings are only achievable when an integrated approach is taken in the design concept, so that BMPs are an integrated part of the development, not an add-on extra facility and consequently an extra cost.

4.8 MONITORING EFFECTIVENESS

As noted at the start of this chapter, a major compilation of monitoring data on the performance of urban BMPs was collated for the USEPA, as part of the support for the implementation of a programme of measures to address diffuse pollution in the USA (USEPA, 1993). That was based on research undertaken since the BMPs idea was developed in the USA during the 1970s (Schueler, 1987). For a pollutant removal summary review table for just ponds and wetlands, Schueler *et al.* (1992) quoted 58 references, and there are 20 pages of references about urban BMPs in the definitive USEPA publication produced in support of the Coastal Zone Act Reauthorization Amendments of 1990 (USEPA, 1993).

The focus of monitoring effort since then has been two fold:

(i) establishing a better rationale for monitoring the effectiveness of individual types of BMPs, providing consistency to permit comparisons between different studies

(ii) assessing the effectiveness of BMPs at a catchment scale.

The Center for Watershed Protection in Maryland, USA, compiled a simple database of 139 BMP pollutant removal performance studies. For inclusion in the database a study had to meet three criteria: a) collect at least five storm samples, b) employ automated equipment for flow or time-based composite samples, and c) have a written documentation of the method used to compute removal efficiency. The database usefully revealed the variation in the types of BMP system that have been studied and the pollutants investigated. For example in 1997 there were no studies on that database for: biofilter, filter/wetland, filter strips, or infiltration basins; only one of bioretention, and only two or three studies for wet swale, infiltration trench, porous pavement, and perimeter sand filter. Table 4.5 shows the pollutants measured by 123 of the studies (Schueler, 2000). It is notable that only nine per cent of the studies measured hydrocarbons (arguably the most ubiquitous toxic

Table 4.5 Frequency that selected stormwater pollutants were monitored in 123 BMP performance studies (Schueler, 2000)

Stormwater parameter	*Percentage of studies in which measured*
Total phosphorus	94
Total suspended solids (TSS)	94
Nitrate-nitrite nitrogen	71
Total zinc	71
Total lead	65
Organic carbon	56
Soluble phosphorus	55
Total nitrogen	54
Total copper*	46
Bacteria	19
Total cadmium*	19
Total dissolved solids	13
Dissolved metals	10
Hydrocarbons	9

[a] Excludes studies where parameter was below detection limits

contaminant of urban run-off), and yet 94 per cent measured total phosphorus, which is good for confidence in the pollutant removal performance of BMPs for the modest number of locations where BMPs must reduce total phosphorus.

Pollutant removal efficiencies for a range of BMPs and pollutants are given in Table 4.6, from Schueler (2000). Whilst these values do give some idea of performance, it is unsafe to draw firm conclusions about the relative effectiveness of the systems. Sometimes influent flows (the untreated drainage) can be relatively unpolluted so there is little scope for large reductions in the concentrations of contaminants. Consequently BMP effectiveness can appear poor, if not checked against concentrations in the discharge. For ponds and stormwater wetlands, sampling at inlet and outlet can produce very misleading results, since the water quality at the outlet from a long residence time system does not have much of a relationship to inlet quality sampled at approximately the same time.

There are also large variations in values measured for different size storm events and the minimum of five events specified by Schueler is insufficient for statistical analysis. These and other issues are considered in detail in Strecker *et al.* (2001) and Clary *et al.* (2001). Roesner (1999b) suggested that renewed efforts should be made to evaluate the effectiveness of urban BMP facilities from a range of states/countries and a variety of hydrological conditions; always specifying the latter as well as the former. Consequently, all those recommendations have been taken forward

Table 4.6a Comparison of median pollutant removal efficiencies (%) reported for selected practice groups (from Schueler 2000)

Practice Groups	*N*	*TSS*	*TP*	*Sol P*	*Total N*	*NO_x*	*Carbon*
Detention Pond	3	7	19	0	5	9	8
Dry ED Pond	6	61	20	−11	31	−2	28
Wet pond	29	79	49	62	32	36	45
Wet ED pond	14	80	55	67	35	63	36
Ponds [a]	**44**	**80**	**51**	**66**	**33**	**43**	**43**
Shallow marsh	23	83	43	29	26	73	18
ED wetland	4	69	39	32	56	35	ND
Pond/wetland	10	71	56	43	19	40	18
Wetlands	**39**	**76**	**49**	**36**	**30**	**67**	**18**
Surface Sand Filters	80	87	59	−17	32	−13	67
Filters[b]	**19**	**86**	**59**	**3**	**38**	**−14**	**54**
Infiltration	**6**	**95**	**70**	**85**	**51**	**82**	**88**
WQ swales [c]	**9**	**81**	**34**	**38**	**84**	**31**	**68**
Ditches	**11**	**31**	**−16**	**−25**	**−9**	**24**	**18**

Notes

N Number of performance monitoring studies. The actual number for a given parameter is likely to be slightly less.

Sol P Soluble phosphorus, as measured as ortho-P, soluble reactive phosphorus or biologically phosphorus.

Total N Total Nitrogen

Carbon Measured of organic carbon (BOD, COD or Toc).

[a] Excludes conventional and dry ED ponds.

[b] Excludes vertical sand filters and vegetated filter strips

[c] Includes biofilters, wet swales and dry swales.

in a new national urban BMPs database, developed in the USA by the American Society of Civil Engineers (ASCE), working with the USEPA to develop a National Stormwater Best Management Practices (BMP) Database (Clary *et al.,* 2001, and web site at www.bmpdatabase.org). Seventy-one studies (at 59 sites) were used in the initial compilation, with scores more to be added, including study sites in Canada and elsewhere. Key categories of data requested for this database were: (1) sponsoring and state agencies, (2) test site location characteristics, (3) watershed characteristics, (4) BMP design and cost criteria, (5) monitoring locations and instrumentation, (6) monitoring costs, (7) precipitation data, (8) flow data, and (9) water quality data. The database is currently being expanded to include data from other countries, including Canada and the UK.

The more rigorous approach of the ASCE/USEPA database does not mean however that other studies have no value; many produce useful information about

Table 4.6b Median stormwater pollutant removal reported for selected practice groups – faecal coliform bacteria, hydrocarbons and selected trace metals (%)[d]

Practice groups	*Bacteria* [e]	*HC* [f]	*Cd*	*Copper*	*Lead*	*Zinc*
Detention & Dry ED Ponds	78	ND	32	26	54	26
Ponds[a]	70	81	50	57	74	66
Stormwater Wetlands	78	85	69	40	68	44
Filters [b]	37	84	68	49	84	88
Infiltration	ND	ND	ND	ND	98	99
Water Quality Swales[c]	(-25)	62	42	51	67	71
Ditches	5	ND	38	14	17	0

[a] Excludes dry ED and conventional detention ponds.
[b] Excludes vertical sand filters and vegetated filter strips.
[c] Includes biofilters, we swales and dry swale.
[d] N is less that 5 for some BMP groups for bacteria, TPH and Cd, and median should be considered provisional.
[e] Bacteria values represent mean removal rates.
[f] HC = hydrocarbons measured as total petroleum hydrocarbons or oil/grease.

Text Box 4.7 The Scottish Universities SUDS Monitoring Project

A loose association of five Scottish universities have been working together to develop and sustain a monitoring programme, funded and steered by the local stakeholder organizations (see Jefferies, 2003). The project produced a national database of sites (nearly 800, and some 3000 individual facilities recorded) that provides information about the type of system most commonly used in Scotland (filter drains), which systems are rarely used (infiltration systems) and the distribution of SUDS geographically (see Wild *et al.,* 2002). The database of SUDS type and details forms the first level of the monitoring strategy of the group: type A sites are simply those listed on the database.

Two additional categories were accepted for the monitoring strategy, recognizing the limitations of resources for monitoring work, and the need for possibility of collecting useful information short of expensive detailed investigations. Thus type B sites are those selected for mainly qualitative observations, e.g. for amenity characteristics and visible evidence of failures.

Type C sites are those relatively few sites selected for detailed measurement of hydrological performance and chemical monitoring of water quality. The leading universities are: Abertay Dundee; Aberdeen,; Edinbugh; Heriot-Watt and Stirling.

Contact c.jefferies@abertay.ac.uk or diffuse.pollution@sepa.org.uk

local circumstances. In Scotland research to investigate the effectiveness of urban BMPs has been undertaken by a group of universities working together to provide a range of expertise, access to research funds and geographic coverage across Scotland (see Text Box 4.7). A summary report of the monitoring programme has been published (Jefferies, 2003). In particular, the Scottish studies have demonstrated the effectiveness of urban BMPs as hydrological controls (Jefferies, 2003; Macdonald 2003). Two of the categories of BMP identified as neglected for monitoring work (swales and filter drains) were investigated in the Scottish research (see below, and Fgures 4.8 and 4.10).

The comparative performance of a conventional tarmacadam car park and permeable block car park in Edinburgh was referred to in Section 4.4. Macdonald (2003) compared the quantities of run-off from conventional tarmacadam surfaces with swales and permeable pavement, and usefully advocates the use of the term 'benefit factor' for describing the values in the comparison. It is a value calculated to summarize the hydraulic benefit gained from installing SUDs (BMP) facilities. It is a volumetric measure expressed as the total volume of SUDS run-off compared

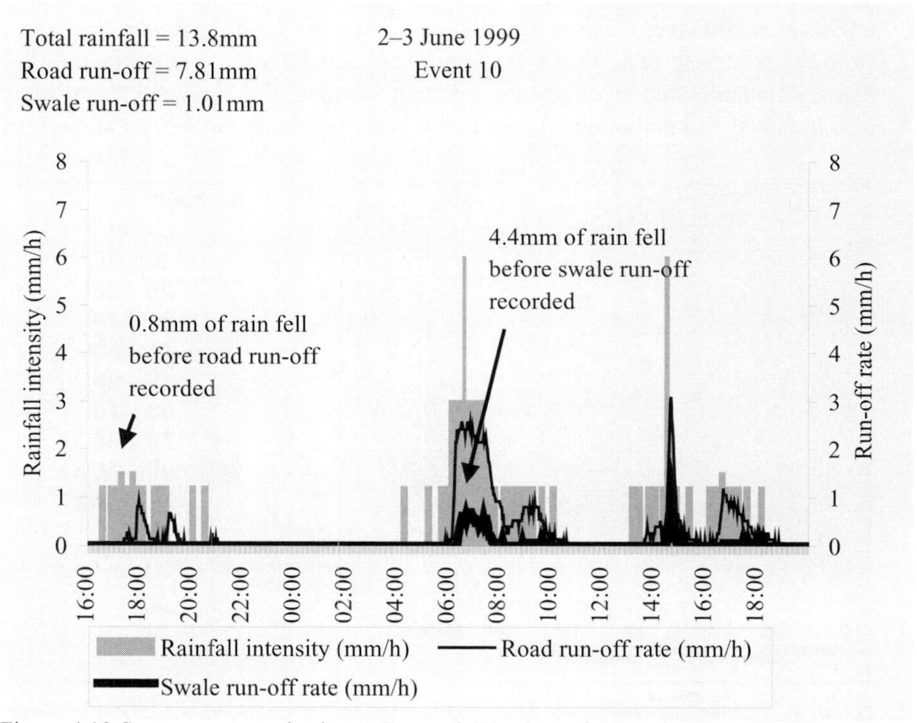

Figure 4.10 Storm event monitoring at Emmock Woods swale, Dundee (Macdonald 2003)

to the run-off from the traditional system, and is calculated using only events that produced run-off. It is also important therefore to report the number of monitored events for which the SUDS retains all the rainfall (see Figures 4.8 and 4.10); the number of events retained. For example, for an event at the porous pavement car park represented in Figure 4.8, when a total run-off of 0.95mm was measured from the porous car park and 4.27mm from the neighbouring tarmac car park, the BF value was 77.8%, i.e. the porous car park produced 77.8% less run-off than the traditional one.

Figure 4.3 shows the performance of two swales in relation to incident rainfall and the frequency and quantity of run-off (from Macdonald, 2003). The frequent failure to produce any discharge after moderate rainfall explains the lack of data from water quality monitoring evident in Table 4.7

The Scottish research has also investigated the contamination of sediments in SUDS ponds (e.g. Heal and Drain, 2003). Figure 4.11 shows example results for nickel contamination in a retention pond in Dunfermline. The monitoring shows that most of the contamination is in the sediments of the retention pond forebay. Comparisons with concentrations in freshwater stream sediments were made in Clarke *et al* 2003 (see Chapter 8, Section 8.5.2, Tables 8.2–8.3). Results measured from the relatively new sites in Scotland were generally less contaminated than heavily contaminated urban stream sediment samples. The Heal and Drain study also included an international literature comparison, involving 396 datapoints, of

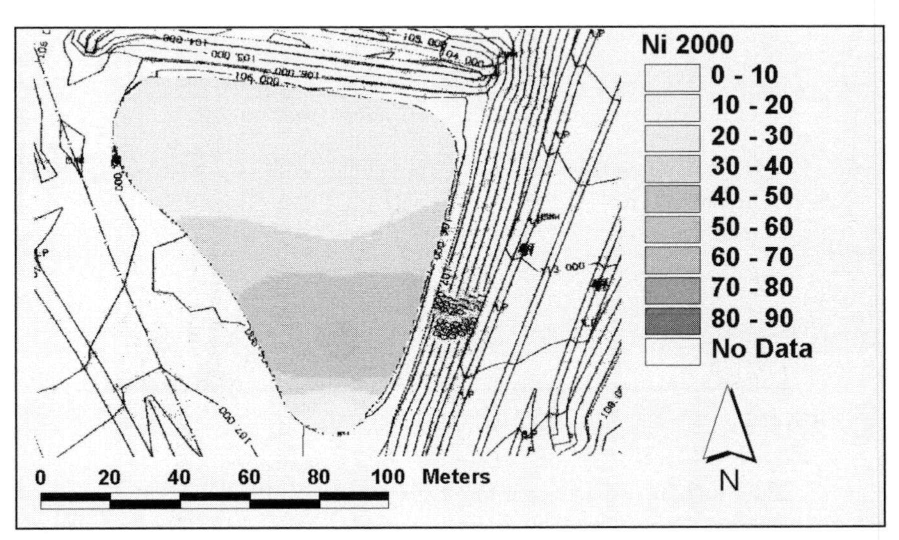

Figure 4.11 Nickel contamination of sediments in Halbeath retention pond, Dunfermline (mg kg^{-1} dry weight) July 2000 (from Heal and Drain, 2003).

Table 4.7 Pollutant removal performance of two permeable car parks (Royal Bank and NATS) and two swales (Emmock Woods and West Grange) in Scotland (based on event mean concentrations). Where no value reported, there was no discharge at time of sampling (from Macdonald and Jefferies, 2003)

	Royal	*NATS Permeable Car Park*			*Emmock Woods Swale*			*West Grange Swale*		
	Bank	*Tarmac*	*Porous*	*+/-*	*Road*	*Swale*	*+/-*	*Road*	*Swale*	*+/-*
TSS (mg/l)	14.9	30	19	32%	1057+	299+	72%	343	96	54%
BOD (mg/l)	2.2	4.8	1.7	49%				5.4	4.5	14%
CU (mg/l)	5.2	5.05	10.9	+25%				28	52	+85%
Ni (mg/l)	1.7	4.64	3.8	63%				6.3	3.1	50%
Zn (mg/l)	22.2	29.5	42	+42%				82.1	93.7	+14%
Hydrocarbons (mg/l)	1.97**	1.07	0.47	69%				1.4	0.9	36%

+/– Values are reductions except where shown positive (+14% indicates increase of zinc in swale)

Tarmac/road are the conventional (non SUDS) surfaces at the respective locations.

* Only one event was sampled at Emmock Woods Swale.

** From one event. Data from a second event was all below detection limit.

which 21 were SUDS. That survey found a wide range of contamination: some stormwater pond sediments were similar to uncontaminated background samples, others were a similar quality to contaminated soils (Heal and Drain, 2003). The authors recommended that a framework for testing SUDS sediments and providing guidance for removal and disposal, as part of routine maintenance is needed.

Effectiveness on a catchment scale is more difficult to evaluate since it requires the retro-fitting of BMPs after pre-assessment of water quality in a watercourse that is primarily influenced by urban drainage (contaminated run-off). In 1999 the US Geological Survey, in co-operation with the State of Vermont and the City of Burlington, with additional support from the Lake Champlain Basin programme, initiated a study of the effectiveness of urban BMPs in improving the water quality of Englesby Brook. An estimated 24 per cent of the catchment of the brook is estimated to be impervious surfaces, land-use is: residential (56 per cent), commercial, industrial, and educational (23 per cent), golf course (18 per cent), forest (3 per cent) and parks and recreation (les than 1 per cent). A seven-year monitoring programme is evaluating stream quality before, during and after BMP implementation (latter commenced in 2001). The USGS is monitoring stream flow, phosphorus, suspended solids, nitrogen, specific conductance, temperature, pH, dissolved oxygen and turbidity. The City of Burlington is monitoring *E. coli* bacteria. (USGS, 2003).

4.9 THE TREATMENT TRAIN CONCEPT IN PRACTICE

The treatment train concept has been championed by Bryan Ellis (e.g. Ellis, 1982), and more recently by Bray (2000) in the UK, and Roesner (1999b) and others in the USA and internationally. As explained in earlier chapters it recognizes that BMPs are most likely to be effective when in series or other combinations. A train of measures progressively reduces pollutant loads. It also provides for management of major pollution risks and incidents closer to source, and thereby protects and maximizes the amenity potential of any regional features such as retention ponds or wetlands. The Watershed Management Institute (1997) note that there is increasing interest in the BMP treatment train concept, wherein several types of stormwater controls are grouped together and integrated into a comprehensive stormwater management system. Table 4.8, from the Watershed Management Institute (1997), with indicative BMPs added, lists stormwater pollutant removal mechanisms for a range of urban run-off pollutants, and also the conditions that promote effective removal. It is evident that a variety of conditions and processes are involved in the removal of pollutants, and in most situations the BMPs will be expected to control more than one class of pollutants. The treatment train concept enables appropriate measures to be incorporated into a stormwater management system in recognition of those facts.

Table 4.8 Summary of stormwater pollutant removal mechanisms: a driver for the treatment train concept (after Watershed Management Institute, 1997, modified to indicate possibly effective BMPs).

Mechanism	*Pollutants affected*	*Removal promoted by*
Physical sedimentation	Solids, BOD, pathogens, particulate COD, P, N, synthetic organics	Low turbulence, increased residence time *Ponds and wetlands*
Filtration	Same as sedimentation	Fine dense herbaceous plants, constructed filters. *Swales and filter strips*
Soil incorporation	All	Medium-fine texture *Swales, filter strips, detention and infiltration basins*
Chemical precipitation	Dissolved P, metals	High alkalinity *Limestone filter drains*
Adsorption	Dissolved P, metals, synthetic organics	High soil Al, Fe, organics, circumneutral pH (around 7) *Infiltration basins, swales or biofilters, depending on soil types*
Ion exchange	Dissolved metals	High soil cation exchange capacity *As for previous case*
Oxidation	COD, petroleum hydrocarbons, synthetic organics	Aerobic conditions *Ext detention, swales, filter strips (for residues degradation), retention ponds and wetlands for waterbody pollutants*
Photolysis	Same as oxidation	Sunlight *As for oxidation*
Volatilization	Volatile petroleum hydrocarbons, synthetic organics	High temperature and air movement *Ext. detention, swales, filter strips.*

continued

Table 4.8 continued

Mechanism	*Pollutants affected*	*Removal promoted by*
Biological microbial decomposition	BOD, COD, petroleum hydrocarbons, synthetic organics	High filter media surface area (including plants) and soil organics *Retention ponds and stormwater wetlands . (Swales, filter strips detention basins for residues).*
Plant uptake and metabolism	P, N, metals	High plant surface area and activity. *Wetlands*
Natural die-off	Pathogens	Sunlight, Plant excretions, saline water *Retention ponds and stormwater wetlands . (Swales, filter strips detention basins for residues). .*

4.9.1 Industrial estates and commercial developments

The idealized diagram in Chapter 3 that indicates selections of facilities that would be sought for draining an industrial estate, (Figure 3.4), is based on an actual site in Wisconsin, USA and featured in the diffuse pollution video of the former IAWQ (now IWA), *Nature's Way*. Other examples include a development currently based on the timber industry near Lockerbie in south-west Scotland. There, heavily contaminated run-off from a de-barking yard is first settled in a concrete settlement chamber and oil interceptor, then passes along swales across the site, picking up site drainage as it goes past additional development areas. The estate drainage finally discharges to a treatment wetland complex, equivalent to a retention pond in capacity terms (see Chapter 3). There are progressive reductions in pollutant concentrations from the de-barking yard area at the head of the drainage network, along the swales to the ponds (see Figure 4.12). The regional facility comprises a first pond with penstock valves at inlet and outlet, to enable the pond and influent drainage to be isolated in the event of a major spill of oil or chemicals on the industrial estate. The second pond provides further capacity for treatment and in turn discharges to a wetland area, prior to discharge into the receiving watercourse. The total storage capacity of the system is sufficient to also meet flood control requirements in respect of the capacity of the small watercourse to receive the run-off from the new estate and its largely impervious surface area. The stormwater wetland has been integrated within a total area of 11 acres, comprising wet grassland, open water, ditches, temporary flashes and perimeter hedgerows. The Royal Society for the Protection

Housekeeping: oil storage tank bunded to contain spills (left).

Source/site controls: roadside swales for road and yard run-off (below).

Regional control: retention pond (left) and wetland (below left).

Inlet to initial sedimentation pond designed to facilitate oil separator function, and outlet from that first stage pond can be closed to contain any serious oil or chemical incidents.

Lastly, wetland receives outflow from pond for final treatment prior to discharge.

Figure 4.12 Stormwater treatment train for an industrial estate in Scotland.

of Birds have prepared a management plan form the site, which has already supported regionally important numbers of lapwing and snipe during 2001.

The treatment train approach is also well demonstrated at two motorway service areas in England, which share many of the drainage characteristics of industrial and commercial estates. The drainage arrangements at one of them, Hopwood Park (M42) near Bromsgrove in the English Midlands, are shown in Figure 4.13 and described in Bray (2000). The site is naturally split into two drainage areas. The larger area receives run-off from coach parking and other hard standing areas, with run-off from small storms passing through a series of ponds providing successive degrees of treatment (amounting in total to a retention pond) but importantly intercepting intermittent shock loads of pollution without overwhelming downstream stages in the treatment train. A by-pass swale runs parallel to the ponds in series, taking run-off from large storm events, to prevent wash out of pollutants from the treatment ponds. The two streams combine at a final (flow control) pond prior to discharge to a small stream with a limited capacity to receive the run-off from the developed area.

Figure 4.14. shows the drainage facilities serving the heavy goods vehicles (HGV) parking area at Hopwood. A series of measures in series provide for management of pollution risks. An interceptor trench collects run-off from the hard standing area and overflows across a grass filter strip. Run-off is subsequently collected in a gravel drain discharging to two ponds in series, prior to discharge to a tributary of the Hopwood stream. Routine monitoring has shown good quality in the final ponds (including biological assessments). Anecdotal indications are that the system is robust and copes well with occasional spills of oil as well as routine background contamination (see Text Box 4.8).

4.9.2 Housing and mixed urban areas

For housing and mixed urban developments, implementation of the treatment train approach has been demonstrated and monitored at the Dunfermline East Expansion site (DEX), (McKissock *et al.*, 2001, and Jefferies, 2003). Spitzer and Jefferies (2003) have undertaken hydrological evaluation of the benefits of a treatment train approach at DEX, involving comparisons between two regional retention ponds, one with no upstream attenuation measures, the other with detention basins for flow control. Difficulties associated with other differences between the two situations complicate the evaluation, but initial findings indicate some flow control benefits and associated savings in size of the facility served by a treatment train. The work has also found evidence of pollution benefits too; oily sediment being retained in detention basins, rather than deposited in retention ponds from which subsequent removal will be more expensive, and detection of wrong connections prior to aesthetic impacts in the retention ponds which are more subject to recreational use. Occasional pollution incidents (e.g. an oil spillage and an illegal discharge of cement washings

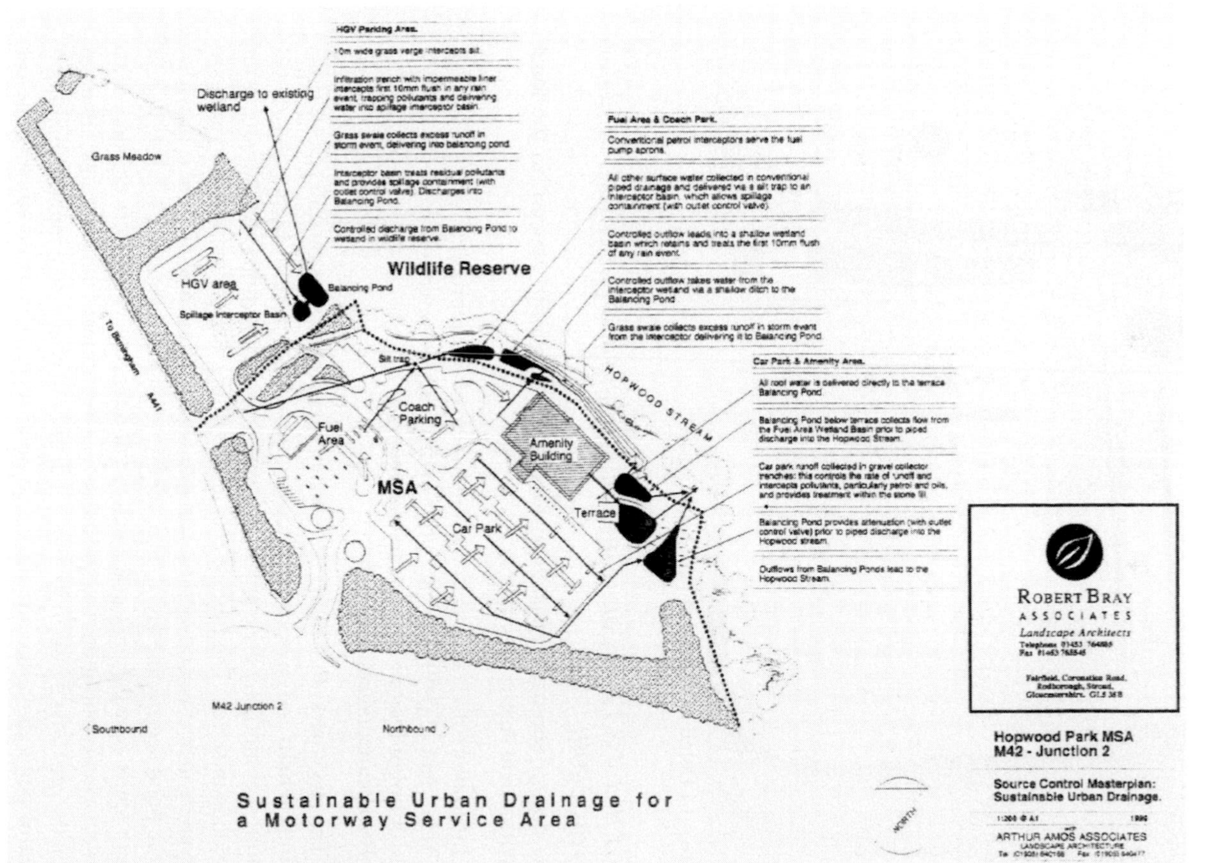

Figure 4.13 Layout of Hopwood Park

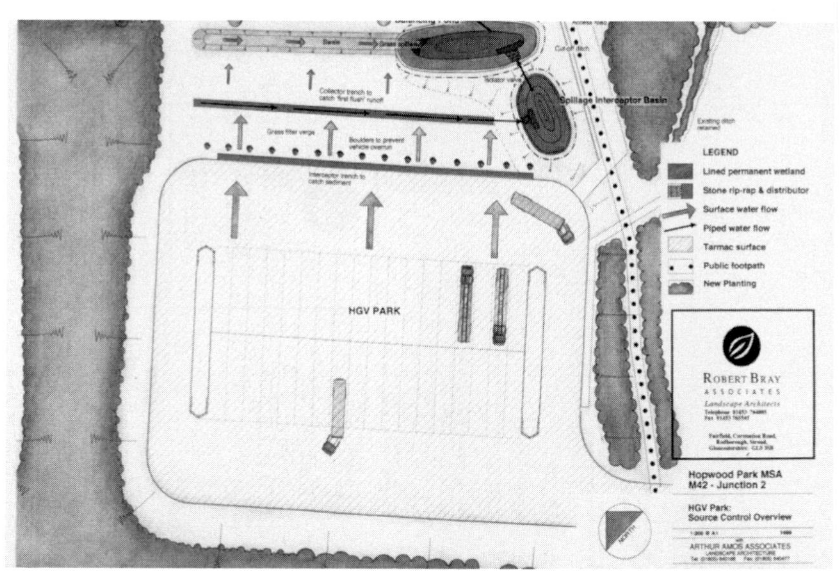

Figure 4.14 Detail of Hopwood Park lorry parking

Text Box 4.8 The treatment train at Hopwood Park MSA, M42. (Bray, 2000)

The design approach considered four areas of priority concern to the local Environment Agency regulators:

- Infiltration of clean water to groundwater where appropriate
- Treatment of polluted water prior to discharge to watercourses
- Control of the rate of flows to watercourses
- Environmental enhancement.

The HGV park, fuel filling area, coach park and service yard all pose serious pollution risks and have an extended treatment train of stormwater management features, comprising grass filter strips and gravel collection (filter) drains, prior to ponds in series. The highest risk areas (filling station and service yard) are served by their own conventional oil interceptors. Capacity of 39m^3 is provided in the pond serving the HGV area, (equivalent to a road tanker load) to accept major spills. The car park and amenity building roof water were considered to present a lesser pollution risk, and are served by shorter treatment trains, although even the roof area drains to its own pond, prior to connection into the final flood control pond above the site discharge point to the Hopwood stream. The car park is served by a sub-surface collector trench treating run-off from the first

10mm of rainfall. A by-pass channel conveys stormwater directly to a pond, which overflows into the main balancing pond serving the site prior to discharge.

The site is operated by the Welcome Break chain of service station operators, who have been so pleased with the drainage system that they have encouraged their customers to walk the drainage system as a nature trail, and have also provided interpretative signs and leaflets, in partnership with the Environment Agency. Full details are given in Bray (2000), and monitoring information in Bray (2001a).

Houseplot swale.

Road swale.

Retention pond.

Figure 4.15 Treatment train for housing development in Illinois.

from a truck) were more evident on the pond with no treatment train (Jefferies, 2003). A comparable treatment train comprising swale, pond and wetland has been established in Illinois for a comparable urban development (see Figure 4.15).

4.9.3 Major highways and motorways

For major trunk roads and motorways, a demonstration of treatment train measures is being studied by SEPA along the M74 in sout-west Scotland (McNeill, pers. com. 2003). Roadside filter drains collect run-off directly from the road surface and then discharge to enhanced extended detention basins, prior to discharge to water courses (typically salmonid rivers and streams of the Annan catchment, described in SEPA 2002). The detention basins were originally provided simply as treatment ponds to protect the receiving waters during the construction phase, successfully intercepting silt disturbed during road building (McNeill, 1998). On completion, (including as a last measure, provision of the roadside filter drains), the silt trap ponds were re-instated as extended detention facilities (McNeill, 2000). Oil, PAHs, suspended solids and toxic metals are the principal pollutants associated with road run-off. Schueler (1987) reported suspended solids removal rates of 75–90 per cent for 'water quality trenches' (filter drains); since much of the metals and hydrocarbons load in run-off is adsorbed onto particulates, that will also remove significant proportion of those contaminants. Research by Pratt *et al.* (1999) has shown that hydrocarbons will breakdown by microbial activity in stone fill, (albeit beneath permeable pavements) provided there is sufficient oxygen diffusing into the system. The filter drains will remove some of the solids and associated metals, and the extended detention basin with permanent pool will provide further removal plus an option to manage gross pollution incidents associated with traffic accidents. Volatilization, photolysis and oxidation (see Table 4.8) will further reduce hydrocarbons left exposed as surface deposits on the sides of the extended detention basins, as water levels drop in between major storm events. The BMPs along the M74 in Scotland are shown in Figure 4.16.

4.9.4 Education and changing behaviours as a BMP– the first stage in a treatment train

Finally, the importance of changing behaviours for the working or urban populations in a catchment, to minimize pollution load for treatment, is the first stage in any stormwater treatment train. Zangbergen (2002) argues that public education is a BMP and can lead to measurable improvements in water quality. It should not be seen as a stand-alone alternative to a best practice drainage system, however, but as part of the BMPs suite to protect the environment Wilson *et al.* (2003) make that point, prompted by considering the degree of contamination of urban stream

Figure 4.16 Highway BMP treatment train, Scotland. (Photos by Neil McLean.) Highway run-off into filter trench (left). Filter trench outflow into detention basion (M74) prior to discharge to stream (right).

sediments by persistent pollutants, most of which are associated with road traffic. For other urban pollutants too there is the same requirement for education efforts to minimize loads if BMPS are to be effective, For example, diffuse sources of faecal indicator organisms are controlled in the first instance by pet owners collecting faeces from their pets, reducing environmental contamination at source (and municipal authorities regulating such behaviour, as well as controlling other sources such as pigeon roosts etc., see O'Keefe *et al.,* 2005). BMPs will be most likely to be effective and sustainable at minimal (but still significant) cost if such efforts to reduce influent loads are recognized as integral to success. As for any other treatment or management technology, overloading will impair performance. Reduced use of cars is another behavioural change that can directly reduce pollutant loadings, but the generation of toxic metals in run-off also needs controlling at source by dialogues with the motor vehicle and construction industries. That and related actions are beyond the scope of conventional BMPs and considered in Chapters 7 and 8.

4.10 BMPS AND BIODIVERSITY

The primary effect of BMPs on biodiversity should, of course, be the consequent prevention of pollution in the water bodies of the catchment, as implied in the preceding section on effectiveness. BMPs can also have effects besides reducing pollutant loads. Roesner (1999b) has argued that the biggest impact of urban run-off in small urban watercourses may be hydrological – the destabilization of stream banks and bed by shock flows off the impervious surfaces of unmitigated urban development. In Europe, the Water Framework Directive will require the remediation of anthropogenic impacts on good ecological status of water bodies: the retro-fitting of BMPs to mitigate the impacts described by Roesner may become an important driver for the technology.

A holistic view of environmental issues needs to be taken for the implementation of BMPs in a catchment. For example BMPs such as ponds can sometimes have a negative impact on trout fisheries in cold streams in warm climates. The potential in warm climates for BMP ponds to warm up receiving streams as a consequence of storing quantities of stormwater in open pools exposed to sunshine for hours is especially an issue for on-line ponds, in regions at the natural southern limit of the geographic range of cold water species. Schueler (1987) reviews the problem and advocates mitigation measures such as deepwater draw-off from ponds. The environmental impact of stormwater ponds is further discussed in Schueler and Galli (1995).

Some BMPs such as ponds and stormwater wetlands can themselves have some wildlife interest, in addition to their primary purpose of protecting the receiving water. Design advice to enhance the wildlife value of such facilities is given in Schueler (1987), Campbell and Ogden (1999), and SEPA (2000a). Primarily from an ornithological stand point, Worrall *et al.* (1997) reviewed the wildlife potential of reed bed treatment systems in the UK, and Knight (1997) reviewed constructed wetlands in relation to wildlife habitat and public use benefits.

The wildlife value of some stormwater ponds in Scotland was evaluated by surveys undertaken for the Scottish Environment Protection Agency, SEPA, by Pond Action (reported with reference to the types of BMPs surveyed, in Walker *et al.,* 2000); summary data are given in Table 4.9. In parallel, invertebrate surveys of retention ponds and a stormwater wetland have also been undertaken on behalf of the developer involved in the Duloch Park part of the Dunfermline DEX site, which is a major UK demonstration site for this technology (see D'Arcy and Roesner, 1999 and McKissock *et al.,* 2001). In comparison with semi-natural unpolluted sites; both datasets have indicated far higher levels of diversity than expected (expectations were sensibly low, based on the presumption that even a low value is higher than would be returned for the tarmacadam or grass lawn alternative uses of the urban land within the developments). That the biodiversity in the relatively few ponds surveyed was moderate to high perhaps reflects the fact that they were primarily large water bodies (also serving as flood control devices in most instances) offering plenty of dilution (settled water) for first flush inputs of pollutants. In several instances they were also protected from the most severe impacts by being at the downstream end of a treatment train of measures. Elsewhere, it is probable that the main opportunity for wildlife value will be in the margins of ponds and wetlands, and in side bays or pools that can easily be created that will be by-passed by shock loads of pollutants (see SEPA, 2000a and Walker *et al.,* 2000).

There can be conflicts between amenity and wildlife interests at urban BMP or SUDS facilities. For example a stormwater wetland or pond has an amenity value as a landscape feature, visible to a greater or lesser extent and similarly at least

Table 4.9 Conservation value of SUDS ponds (from Pond Action report as quoted in Walker *et al.*, 2000)

	Motorola Motorway	*Motorola Lower*	*Freeport Upper*	*Houston Caw Burn*	*Dex Calais Wood Marsh*
Invertebrates					
Number of species	40	37	58	24	40
Number of uncommon species	0	1	1	0	0
Conservation value	High	High	Very High	Moderate	High
Plants					
Number of native species	17	12	24	13	25
Number of uncommon species	3	2	1	0	4
Conservation value	Moderate	Moderate	High	Moderate	High

partly accessible to the local population and visitors to the area. If the pond is natural or semi-natural it may have an importance locally for biodiversity too. If the pond becomes a centre for feeding waterfowl and the numbers of the latter grow to unnatural levels, the waterfowl will destroy the vegetation, over-enrich the water with nutrients and enrich it with faecal indicator organisms, FIOs. That would probably eliminate any benefits for example anticipated in relation to removal of urban diffuse sources of FIOs from the urban catchment (a problem described, for example, in O'Keefe *et al.*, 2005).

The conflict between formal tidiness and a wilder look has already been referred to in the discussion on maintenance in section 4.3; it is especially important for wildlife interest as well as costs. Even where stormwater management facilities are themselves of limited biodiversity value, they can form links of sub-optimal natural habitats between the preferred more natural habitats of species. For example water voles (*Arvicola terrestris,* See figure 4.17) European rodents that have suffered a major and sustained decline in the UK, have a survival strategy based on population expansion at a site, followed by dispersal.

In a region with plenty of water voles, each suitable site is likely to be recolonized by dispersing animals from neighbouring sites, so genetic heterogeneity is maintained as well as geographic coverage of an area. But the voles are unlikely to recolonize sites once the distance between them becomes too great. In a study in north-east Scotland for example, Lambin *et al.* (2000) found that no recolonized areas were more than 1.1km from their nearest neighbouring colonies. Lambin *et al.* concluded that colonies in dense clusters are less likely to become extinct. BMP facilities, together with stream restoration programmes (see guidelines in Strachan, 1998 and SEPA, 2000a and 2000b) could be important providers of additional habitats to

Figure 4.17 Water vole *Arvicola terrestris.* A UK biodiversity action plan species sometimes frequenting stormwater ponds. (Photo: Rob Strachan.)

maintain connectivity between remnant populations. Water voles have been found at three SUDS (urban BMP) sites in Scotland, although they subsequently disappeared (hopefully temporarily) from two of them (D'Arcy, pers. com. 2002).

Similarly, another UK threatened species, the great crested newt (*Triturus cristatus*, see Figure 4.18), is also dependent on the close proximity of breeding sites, in order to sustain viable populations. The UK biodiversity action plan (Langton *et al.,* 2001) recommends consideration be given to stocking suitable ponds as part of a translocation or re-introduction programme. BMP retention ponds and stormwater wetlands – especially if protected by a treatment train of measures (see Section 4.9) – should be suitable.

The survival of newts also raises another conflict area between biodiversity and recreational use of amenity ponds. Angling is a major leisure activity and local populations in the vicinity of urban ponds will not be slow to try and stock the ponds with fish. That can be disastrous for the successful colonization by, and development of viable populations of, amphibians in the ponds. Irrespective of regulations and educational campaigns, fish stocking is likely to happen sooner rather than later (anecdotal evidence at DEX and from anglers comments, see also the review of the distribution of freshwater species of fish in the British Isles, Maitland and Campbell, 1992). It may therefore be sensible to stock (with the prior consent and approval of the appropriate nature conservation bodies) new ponds in urban areas with amphibians as soon as possible, in the hope that viable populations can develop there and sufficient aquatic vegetation, prior to likely, albeit undesirable and possibly illegal, fish stocking.

In conclusion, the conservation of biodiversity is a challenge that requires interdisciplinary working by ecologists with engineers, planners, regulators, and developers. It will also require engagement with the general public if opportunities are to be realized. It will not be achieved by one group in academic isolation believing it holds all the answers.

Figure 4.18 Great crested newt (*Triturus cristatus*), an occasional inhabitant of stormwater ponds in the UK. (Photo: Froglife www.froglife.org.uk)

Less mowing & use of native vegetation prevents soil erosion and nutrient losses, and provides more opportunities for wildlife.

Severe grass-cutting regimes minimize potential for wildlife in and around the feature, as well as (in this example) adding nutrients to the pond in the form of grass-cuttings and fertiliser leachate.

Treatment train: West Bend Industrial Park South, Wisconsin.

Plot edge swale.

Roadside swale.

Inlet to regional stormwater wetland.

The wetland.

5

Best management practice for agriculture

5.1 POLLUTION CONTROL STANDARDS

Pollution has typically been controlled in the past by a system of legally enforceable discharge consents to which testable conditions may be attached. The level of discharge of pollutants has been set by reference to environmental quality standards (EQS), which indicate the maximum pollutant concentrations consistent with appropriate use of the watercourse.

Such a system is unworkable in the case of agriculturally derived diffuse pollutants (D'Arcy and Frost, 2001). There is no point source to consent or monitor. The sources of the pollutants are rarely well defined and the pathways of movement of the pollutants are frequently poorly understood. There is often also no well-defined EQS to act as a yardstick. For example, a river in spate may naturally carry very high levels of suspended solids derived from bank and bed erosion. These levels may be far higher than would be desirable were the suspended solids to be derived from agricultural run-off.

5.2 BEST MANAGEMENT PRACTICE

The best management practice (BMP) approach to the solution to the problems of agriculturally derived diffuse pollution does not seek to set numerical quality targets for individual farmers to meet. Rather, it sets forth a series of good practices which the farmer is encouraged to adopt.

Such practices may be procedural or may be structural. Examples of procedural BMPs include, for example, the use of the optimal rate of the right fertilizer at the right time for any crop, or the adoption of pesticide spraying techniques that prevent spray drift into watercourses adjacent to the crop being sprayed. Many procedural BMPs are simply examples of what traditionally was termed 'good husbandry'. Examples of structural BMPs include the provision of adequate on farm storage for cattle slurry to ensure that the slurry can be applied to fields at the optimal time and in optimal conditions, or the construction of diversion terraces to minimize soil loss resulting from erosion by water.

Not all agricultural environmental improvements are necessarily applied on the farm. The formulation and production of pesticides, which are more specific in their action or leave fewer undesirable breakdown residues in the soil, is undoubtedly also a tool in minimizing the impact of agriculture on the environment but, because it is done in the factory rather than on the farm, it is not classed here as an agricultural BMP.

When a BMP approach is adopted, if numerical targets are set at all, they are set for a catchment as a whole and used as a yardstick to assess the success of the whole scheme rather than the performance of individual farmers. The criterion for judging the performance of individual farmers is rather the extent and the diligence of their application of BMPs to their farming system.

5.3 AGRICULTURALLY DERIVED DIFFUSE POLLUTION

Pollutants fall into five main groups. These are nitrogen, phosphorus, pesticides, suspended solids and micro-organisms (Frost, 1999). Unfortunately all have different sources, different pathways and therefore different potential control methods. Agriculturally derived diffuse pollutants cannot therefore safely be lumped as a single group.

A discussion of the nature of the five main types of pollutant and possible BMPs aimed at reducing their impacts follows.

5.4 GENERAL

Before pollution of water can occur, the pollutants must move from the land to the water. There are a number of possible routes.

It may happen by direct input. Fertilizers or pesticides may be spread too close to watercourses and a proportion of the product may be thrown by the spreader directly into the water. In windy conditions, fertilizers or pesticides may be blown into watercourses. Some agrochemicals are volatile and the product may drift downwind as a vapour to enter water. Animals may have direct access to water for drinking and may defecate or urinate directly into the water.

Pollution may occur by surface run-off from soil. If the rate of rainfall exceeds the infiltration capacity of the soil, or if the whole soil profile above impermeable lower layers is saturated, water will run off across the surface of the soil possibly carrying soil and other pollutants with it. This is particularly likely to occur if the soil structure at the surface has been damaged and the soil is capped with a thin but relatively impermeable layer. The presence of a plough pan or other soil structure damage at depth can have a similar result.

Pollution may occur by leaching. In most agricultural areas, neither rainfall nor evapotranspiration are constant throughout the year. At times when evapotranspiration (the loss of water from soil to the atmosphere by the combined processes of evaporation of water from the soil and transpiration of water by the crop) exceeds rainfall, a soil water deficit will build up and any rainfall will merely reduce that deficit, and will tend to be retained within the soil. When rainfall exceeds evapotranspiration, the soil water deficit will disappear and the soil, said to be at field capacity, can hold no more water. Any rainfall in these circumstances which enters the soil will displace an equal volume of water from the soil. This displaced water may leave the soil via underlying permeable strata to the ground water table, may move horizontally above impermeable underlying strata to the nearest water course or may leave via agricultural field drains. It may carry pollutants with it, in solution, in suspension, or in a colloidal form.

Some pollutants, such as nitrogen, are most likely to be lost from the soil by leaching; others, such as suspended solids, are most likely to be lost by surface run-off.

5.5 NITROGEN

Nitrogen is an essential plant nutrient. In a modern system of agriculture making significant use of artificial fertilizer, nitrogen is likely to produce the largest yield response of any fertilizer used. Sources of nitrogen in soil include nitrogen derived from artificial fertilizer, nitrogen derived from animal manures, nitrogen derived from mineralization of nitrogen-containing compounds in soil organic matter and plant residues, and nitrogen derived from atmospheric deposition.

While elemental nitrogen is clearly very common, comprising about two-thirds of the atmosphere, plants cannot use nitrogen in this form. Before plants can use elemental nitrogen it must be converted into nitrate or ammonium ions (NO_3^- or

NH_4^+). This process is known as fixing nitrogen. Nitrogen is fixed from atmospheric elemental nitrogen by chemical engineering processes in fertilizer factories. It is fixed by microbial activity in soil either by free-living soil algae or by bacteria living symbiotically in the roots of plants, especially legumes. Some is also fixed within the atmosphere by the action of lightning on atmospheric nitrogen. Before the advent of artificial chemical nitrogen, almost all plant (and animal) nitrogen came ultimately from microbial or lightning fixation and this was a major constraint to the productivity of agriculture.

A large amount of nitrogen is held in most soils in an organic form. This nitrogen is a constituent part of soil organic matter, sometimes referred to as humus. Soil organic matter is the remains of plant and microbial residues accumulated over centuries. Typically, topsoil will contain between about one per cent and ten per cent of organic matter although the range is even greater than this. Levels tend to be high in cooler moister climates and under cropping regimes such as grass, and tend to be low in hotter dry climates and under cropping regimes with a high proportion of crops such as vegetables, maize and potatoes. Typically, soils will contain between 1.5 and 15 tonnes per hectare of nitrogen in an organic form. During the growing season, as the soil warms up in summer, a proportion of this pool of nitrogen will be mineralized, that is converted within the soil to a nitrate or ammoniacal form, and may be taken up by the growing crop. In some circumstances, crops may derive almost all their nitrogen from this source. Mineralized soil organic nitrogen is particularly useful in crops, which do most of their growing at times when the soil is warm. It, therefore, is of more benefit to spring-sown crops than to autumn-sown crops. Very large amounts of nitrogen may be released in circumstances where there has been a major change in cropping from a regime which tends to conserve soil organic matter (such as pasture) to one which tends to encourage mineralization (such as arable agriculture using the mouldboard plough as the primary cultivator). After harvest, soils are frequently still warm and mineralized soil organic nitrogen can be a significant source of leached nitrogen in watercourses.

Farmers have always used animal manures as a source of nitrogen for crops. The nitrogen in the manures is derived from the food the animal eats, either grown on the farm or brought into the farm. Ultimately the nitrogen derives from one of the fixation processes described above. Most manures contain nitrogen in a mineral form, usually ammonium ions or urea which is readily converted to ammonium, and in various organic forms which must go through similar mineralization processes as nitrogen derived from soil organic matter described above. In stock-rearing systems where most of the animal food is grown on farm, animal manures are likely to be used effectively and the nitrogen in them to be recycled through the fodder crops. However, in intensive systems where animal food is imported into the farm in large quantities, far more nitrogen may be available in the manures than can possibly be taken up by the crops (including grass) grown on the farm. In these circumstances, environmental problems are likely to result unless effective steps are taken to prevent them. For example, on a west of Scotland dairy farm of about 60ha with 100 milking cows, the annual input of fertilizer nitrogen was 7500 kg. The annual input of nitrogen in feed for the cattle was 8400kg. Off-takes of nitrogen in the form of milk and sold stock amounted to only 2700kg giving an excess of inputs over off-takes equivalent to 220kg/ha.

Ultimately, whatever source nitrogen comes from, it must be converted to nitrate or ammonium ions before plants can use it. Ammonium ions themselves are rapidly converted to nitrate ions within the soil by a microbial process known as nitrification. Nitrate ions are highly soluble in water and will move freely through soil along with water. They are therefore readily leached in circumstances where there is excess rainfall, that is where the rainfall on the land exceeds the evapotranspiration of water from the land. The various organic forms of nitrogen, whether in the form of soil organic matter or in animal manures are protected from leaching as the nitrogen is not soluble. Ammoniacal nitrogen is also protected somewhat from leaching as the positively charged ions are held on the cation exchange sites on clay particles. However, before nitrogen can be used, it will generally be converted to the nitrate form and, whatever its original form, will be susceptible to leaching. The nitrogen cycle is illustrated in Figure 5.1.

Nitrogen pollution will thus generally be by leaching, but pollution by surface run-off also occurs when fertilizers, either chemical or organic, are washed from the soil surface before they are incorporated into the soil. Direct inputs also undoubtedly also occur as a result of poor fertilizer spreading technique.

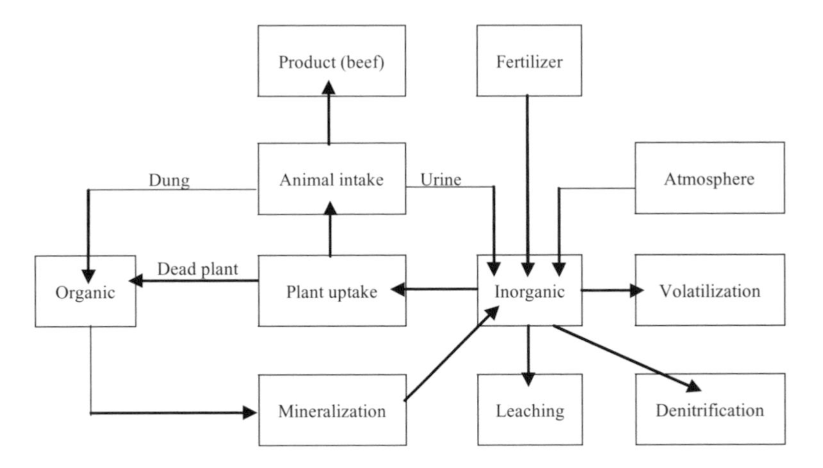

Figure 5.1 Nitrogen cycle

5.5.1 BMPs to control nitrogen pollution

BMPs to control nitrogen pollution are either aimed at ensuring that as much as possible of the nitrogen available for crop use is taken up by the crop or they are aimed at ensuring excess nitrogen is not leached, but is held within the soil for subsequent uptake by following crops.

All the following can, in the correct circumstances, reduce nitrate leaching and water pollution. Not all are applicable in all circumstances.

Correct rate of application based on likely crop uptake

Assessing the correct rate of nitrogen for any given crop is not easy. To some extent, it is necessary to guess what the weather is likely to be like during the season because the rates both of mineralization of soil nitrogen and, in wet years, of leaching of nitrogen during the growing season depend on the weather. Techniques such as incubation and analysis of soil nitrogen and split dressings of fertilizer will help.

Correct timing of application

Similar remarks apply. Too much, too early risks losing nitrogen by leaching (or denitrification) if the weather turns wet, too little too late risks reduced crop yields.

Accurate fertilizer spreading using modern spreaders

Great improvements have been made in fertilizer spreader design making uniform accurate application of fertilizer onto the crop and not onto the surrounding land or water easier.

Improved use of nitrogen in animal manures

(Chambers *et al.*, 2000) Too often, the nitrogen present in animal manures is not fully allowed for in deciding the correct rate of chemical fertilizer to apply. Also, if manures are applied in autumn or winter much of the readily available mineral nitrogen present will be lost to leaching or denitrification.

Account taken of soil derived nitrogen

(Withers and Sylvester-Bradley, 1999) Late-sown crops, such as turnips, can derive most of their nitrogen from mineralized soil nitrogen. The contribution even to autumn-sown crops can be significant. If this nitrogen is not allowed for in deciding fertilizer inputs, it will be lost by leaching. Techniques such as incubation and analysis of soil nitrogen help to predict releases.

Precision farming techniques

Global positioning system (GPS) technology is being increasingly applied in agriculture with detailed maps of within-field variation in soil chemical fertility being compared to maps of crop yield produced by combine harvesters equipped with GPS and yield meters. The result could be more precise determination of fertilizer requirements with the rate being varied according to the yield potential of the area of the field. This may result in less nitrogen leaching.

Use of nitrification inhibitors

Nitrification inhibitors delay the transformation of ammonium ions into nitrate ions. As ammonium ions are protected from leaching by chemically active clay surfaces, nitrification inhibitors added to animal manures can delay nitrogen leaching until the plant has time to take up the nitrogen.

Growing of green manures post harvest

(Christian *et al.*, 1992; Thomsen and Christensen, 1999) After harvest in temperate climates, the soils are still warm enough for mineralization of soil organic nitrogen to continue. If there is no crop to take this nitrogen up, it will accumulate in the soil to be leached out by autumn rain once the soils have regained field capacity. Growing a green manure crop protects this nitrogen by converting it to organic forms which will become available after the green manure is ploughed in prior to sowing the next crop. Care is needed however. In northern latitudes such as in Scotland, the nitrogen produced by the increased rate of mineralization caused by the cultivations necessary to establish the green manure is often not fully taken up by the slow-growing green manure, resulting in increased nitrogen leaching. Local knowledge is often essential.

Early sowing of autumn-sown crops

(Lord and Shepherd, 1996) Sowing autumn crops early ensures that enough growth occurs to take up the soil nitrogen becoming available in the still warm soils. However, the environmental gain of reduced nitrogen leaching must be offset against the frequently increased requirement for pesticides.

Application of high carbon-to-nitrogen mulches

(Vinten *et al.,* 1998) If organic mulches (such as paper-mill waste) with a carbon-to-nitrogen ratio in excess of about 20:1 are added to soil post harvest, any available nitrogen in the soil is locked up by microbes which require a nitrogen source in order to fully exploit the available carbon. The nitrogen will be added to the soil organic matter pool and become available to plants in future years.

Proper scheduling of irrigation water

(Groves and Bailey, 1997) Over-application of irrigation water causes wet soil conditions where leaching can occur even in mid summer. Relating water application to calculated or measured soil moisture deficits can avoid such losses.

5.6 PHOSPHORUS

Phosphorus is also an essential plant nutrient. It was the first chemical fertilizer to be used in the nineteenth century and its use greatly increased crop yields in the many soils which naturally suffer from phosphate deficiency.

Phosphorus in soil is naturally derived from the weathering of the parent material rock from which the soil is formed. If that rock is low in phosphorus, the soil will also tend to be low in phosphorus.

Phosphorus in chemical fertilizers is derived from phosphorus-containing rocks which occur in a number of places throughout the world. The rock is mined and is either ground up and applied directly (rock phosphates) or, more commonly, chemically treated to make a water-soluble chemical fertilizer. Phosphorus fertilizer therefore is got by mining a non-renewable resource. In this respect, it differs from nitrogen, which is so abundant as to be virtually inexhaustible (although the same may not be said of the energy required to fix it).

Phosphorus within soil, whether naturally occurring or added, is almost always adsorbed onto the clay and the organic matter within the soil. Phosphorus in this state is neither available for plant uptake nor for leaching, but it may, of course, still pollute water if the soil itself is eroded and enters the water. A small amount of phosphorus is also present in soil solution. It is this phosphorus in solution, which may be taken up or leached. There is a dynamic equilibrium between adsorbed phosphorus and phosphorus in solution (Figure 5.2). If the phosphorus content of the soil solution falls, either because of plant uptake or because rainwater dilutes the soil solution, phosphorus will tend to desorb from the soil to maintain the concentration of phosphorus in solution. Conversely, if the concentration of phosphorus in the soil solution rises, because, for example, the farmer has added a water-soluble fertilizer, phosphorus will tend to be adsorbed onto the soil and removed from solution.

The quantity of phosphorus in solution at any one time is small, only about 0.2 to 2.0kg P_2O_5/ha. The typical crop requirement of about 30 to 70 kgP_2O_5/ha/annum is mostly supplied by movement of labile phosphorus into solution and thence taken up by the roots. Because so little is in solution at any one time, from an agronomic point of view, phosphorus has always been considered non-mobile and not susceptible to leaching. However, from an environmental point of view, this is not true. The few kilograms per hectare, which might leach from soil, may be of no

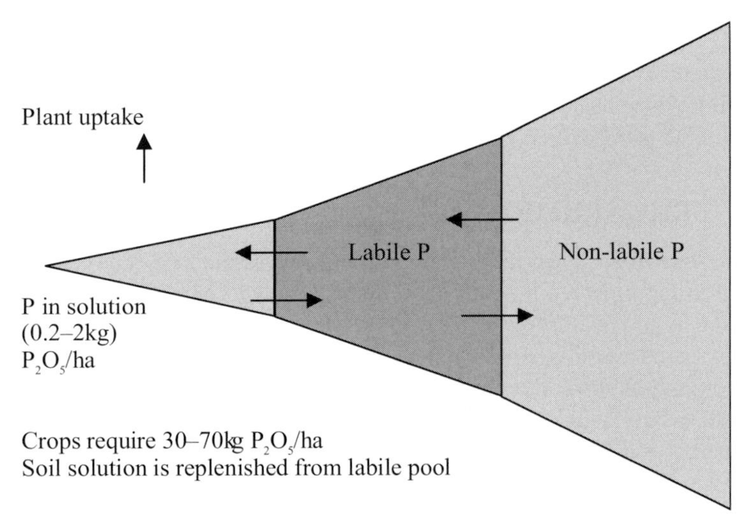

Figure 5.2 Phosphorus partition in soil.

agronomic significance but it may still be sufficient to cause severe eutrophication of sensitive receiving water. Large reserves of phosphorus within the soil, i.e. high chemical fertility, lead to higher concentrations of phosphorus in the soil solution and consequently more phosphorus being leached. There is evidence that the relationship between phosphorus concentration in soil and phosphorus leaching is decidedly non-linear (Figure 5.3).

It will be seen from Figure 5.3 that a phosphorus concentration is reached where losses suddenly become much larger. Unfortunately, the level at which this occurs appears to vary greatly between soil types and no general guidance can yet be given as to environmentally safe phosphorus levels for a range of soils.

Phosphorus pollution may thus occur by run-off, with the phosphorus adsorbed onto eroded soil or chemical and organic fertilizer washed from the soil surface before it has been incorporated. Leaching also occurs, particularly from soils with an elevated phosphorus content, as does direct input where fertilizer spreading practices are poor. Various routes of movement of phosphorus from soil to water along with various possible control measures are illustrated in Figure 5.4.

5.6.1 BMPs to control phosphorus pollution

All the following can, in the correct circumstances, reduce phosphorus transfer from field to water. Not all are applicable in all circumstances.

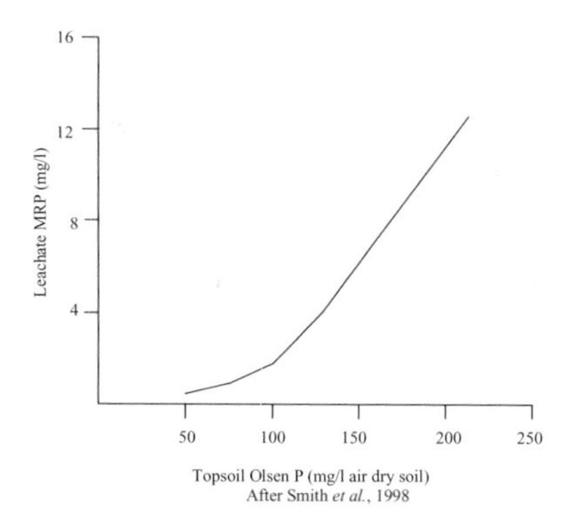

Figure 5.3 Phosphorus leaching as a function of soil content.

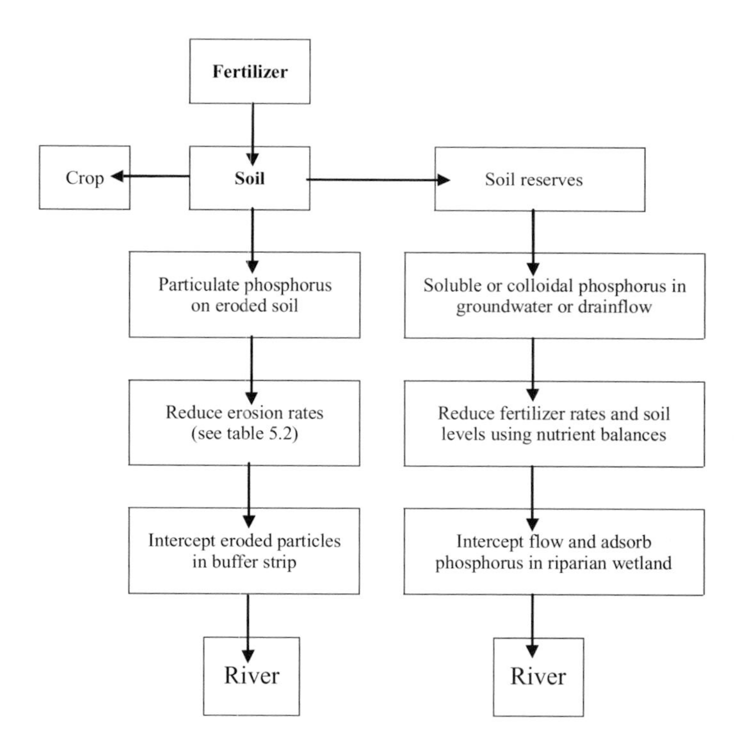

Figure 5.4 Pathways for phosphorus enrichment of waterbodies.

Phosphorus fertilizer rates based on a balance sheet approach

Because the leaching of phosphorus is slow in most soils, the soil can be regarded as a 'bank' for phosphorus. Deposits to the bank are the fertilizer and organic manure applications; withdrawals are the off-takes in the harvested part of the crop. If deposits equal withdrawals then the 'balance', the phosphorus status of the soil, will remain constant. This approach to deciding phosphorus fertilizer rates prevents undesirably high phosphorus levels from building up in soil over a number of years.

Accurate fertilizer spreading using modern spreaders

Great improvements have been made in fertilizer spreader design, making uniform accurate application of fertilizer onto the crop and not onto the surrounding land or water easier.

Account taken of phosphorus in animal manures

(Haygarth *et al.*, 1998) Many farmers in the past have not taken proper account of phosphorus in animal manures. Doing so can greatly reduce or eliminate the need for additional chemical fertilizer and thus prevent the build-up of undesirably high soil phosphorus levels.

Precision farming techniques

Global positioning system (GPS) technology is being increasingly applied in agriculture with detailed maps of within field variation in soil chemical fertility being compared to maps of crop yield produced by combine harvesters equipped with GPS and yield meters. The result could be more precise determination of fertilizer requirements with the rate being varied according to the yield potential of the area of the field. This may result in overall reductions in phosphorus inputs.

Having sufficient land for animal manures

Intensive livestock enterprises, which utilize large amounts of bought-in feed often produce more phosphorus than can be taken up by the crops grown on the farm. Unless the manure is exported from the farm and possibly the area, undesirable accumulation of phosphorus in soil and consequent leaching will occur.

Control of soil erosion (see below)

(Edwards and Withers, 1998) When eroded soil enters water, it carries its attendant phosphorus with it. In certain conditions, particularly if the sediments become anaerobic, the phosphorus can later be released to the water causing eutrophication.

Use of buffer strips

(Heathwaite *et al.,* 1998) Buffer strips, that is linear areas of uncropped land normally, but not necessarily, adjacent to watercourses, may protect that water from inputs of phosphorus either by trapping eroded soils before they enter the water, or by adsorbing dissolved or colloidal phosphorus as it passes across or through the buffer strip. They are also effective at preventing direct inputs by keeping fertilizer spreaders away from water.

Use of river margin wetlands

(Chambers *et al.,* 1993) Special wetlands, created adjacent to watercourses are considered to be more likely to trap phosphorus than dry buffer strips and also offer the possibility of intercepting drain flow.

Manipulation of animal diets to reduce phosphorus input (phytase treatment)

The phosphorus content of animal manures can be reduced by reducing the phosphorus content of the feed which is also treated with the enzyme phytase which effectively increases the animals' efficiency of utilization of the available phosphorus.

Treatment of animal manures to reduce phosphate availability (alum treatment)

(Moore *et al.,* 1998) Treating animal manures with chemicals such as alum or iron oxide reduces the availability of the phosphorus in the manure by causing it to combine in insoluble compounds.

5.7 PESTICIDES

A wide range of chemicals is used by farmers to control pests. These include insecticides, herbicides and fungicides. Some are water soluble and will move in soils as nitrates do. Some are adsorbed onto clay or organic matter and will tend to behave more like phosphorus. Some are volatile and, in certain climatic conditions, may move through the atmosphere to other locations. Most are nowadays designed to break down in soil reasonably quickly. In such cases the toxicity and likely fate of the breakdown products should also be considered.

Because of the range of products involved, no general comments can be made. Great strides have been made by agrochemical companies in recent years to reduce the persistence of their products in the environment or to make their toxicity more specific to the target organism. Not all such changes have necessarily had environmentally beneficial results. For example, the recent change in sheep-dipping chemicals from organo-phosphates to synthetic pyrethroids has produced benefits for the operators of the sheep dips in terms of greatly reduced toxicity at the expense of greatly increased toxicity to invertebrates in rivers which may be polluted by the chemical.

Further progress has been made in the breeding of crop varieties resistant to pests and diseases. It is likely that much further progress will be made in this area.

Pollution by pesticides may occur by leaching of soluble product. It may also occur by surface run-off either washing product into the water directly or moving soil carrying adsorbed pesticides. Direct inputs also occur.

Many pollution incidents however occur not when the pesticides are actually being used according to the manufacturer's recommendations but during handling of the product prior to or after use. Spills in the farmyard during the filling of sprayer tanks (Harris *et al.*, 1991) or losses when spraying equipment is cleaned (Higginbotham *et al.,* 1999) are both common. Modern handing techniques are available to help reduce both these risks.

5.7.1 BMPs to control pesticide pollution

Accurate application

Ensuring that the chemical is delivered to the crop at the correct rate and not to the surrounding area reduces pollution.

Use of modern spraying machinery

Improvements in sprayer design enable the correct dose to be applied to the crop and reduce the risk of spray drift etc.

Targeted pest control techniques

Involves careful monitoring of the build-up of pests to ensure that pesticides are not used before they are really needed.

Integrated pest management

(US Congress, Office of Technology Assessment, 1990) Systems have been developed to farm in such a way that natural predators are used as far as possible to control crop pests and in such a way that the build-up of crop pests is discouraged.

Use of anti-drift agents

These are chemicals which may be added to the sprayer tank mix designed to reduce the risk of spray drift.

Use of spray adjuvants

These are chemicals which may be added to the sprayer tank mix to improve the spreading, sticking or penetration of the pesticide and make it more effective at a lower dose.

Novel seed treatments

Chemicals targeted at seed-borne problems can be dressed onto the seed in controlled conditions avoiding field application of chemicals. Some other crop problems can also be amenable to this approach.

Control of soil erosion (see below)

When eroded soil enters water, it carries its adsorbed pesticides with it.

Buffer strips

(Patty *et al.*, 1995) Buffer strips, that is linear areas of uncropped land normally, but not necessarily, adjacent to water courses may protect that water from inputs of pesticides either by trapping eroded soils before they enter the water or by adsorbing dissolved pesticide as it passes across or through the buffer strip. They are also effective at preventing direct inputs by keeping sprayers away from water.

Good practice in sheep dipping

In sheep farming areas, sheep are often dipped in insecticide to control diseases. This can give rise not only to a range of point source problems, but also to diffuse problems if the sheep are allowed to leave the controlled area before the dip has drained from their fleeces. Some modern pyrethroid dips are extremely toxic to fresh water invertebrates. The use of sheep showers rather than dips can greatly reduce the surplus chemical to be disposed of.

Use of 'biobeds' while filling sprayer tanks

A safer surface than the traditional concrete has been developed for filling sprayer tanks known as 'biobeds'. (Torstensson and de Pilar Castillo, 1997). These involve a pit with layers of organic materials such as straw or peat which encourage the bacterial breakdown of any spillage.

Proper disposal of excess chemical and tank washings

Excess chemical in a sprayer tank or the washings from a sprayer tank can pose serious environmental threats if disposed of incorrectly.

Low volume sprayer washing

Specially designed high-pressure water cleaners have been designed which reduce the volume of effluent to be disposed of when sprayer tanks are washed out.

5.8 SUSPENDED SOLIDS

A great deal of scientific and extension work has been carried on over the past seventy years to reduce the losses of soil from land by soil erosion. Almost all of this work has been aimed at reducing the damage caused to the land rather than the damage caused to the receiving water. Indeed, it is only relatively recently that suspended solids, in their own right as opposed to the pesticides and phosphorus they may carry, have been regarded as a significant environmental pollutant.

Because of this long history of study, an impressive body of literature exists on techniques to control soil erosion. A certain degree of care must be exercised however. Soil erosion at rates of one or two tonnes per hectare per annum may not present any threat to agricultural productivity, even in the long term. Such rates have tended to be regarded as targets to be reached by erosion control. The concept of soil loss tolerance has been widely used. However, such rates may be unacceptably high for sensitive catchments and may result in eutrophication by phosphorus enrichment.

For erosion to occur, the impact of raindrops on soil must break down the larger soil aggregates to produce particles of a size which can be transported by the water. For pollution to occur, there must also be surface flow of water across the soil to carry the detached soil particles to the nearest water course. The rate at which soil will be transported to water depends on both detachment rates and transport rates, the lower of the two being the controlling factor. Most erosion control techniques aim to limit one or the other. For example, stubble mulch techniques (see below) aim to reduce the ability of falling raindrops to disrupt surface aggregates and thus control the detachment rate. Diversion terraces (see below) aim to reduce the volume and velocity of surface run-off water flowing over the land and thus control the transport rate.

Significant soil erosion may occur without water pollution if the eroded material is all deposited before it reaches the water. It should, however, be noted that while larger particles such as sand grains settle out readily from slowly moving water, neither clay particles nor organic matter will do so. Many clay particles are small enough to be kept in suspension by Brownian motion even in still water and organic matter generally floats. Unfortunately, most of the pollutants in the moving soil, i.e. adsorbed phosphorus and pesticides, will be carried on these clay and organic matter fractions of the soil.

Clearly, virtually all pollution of water by suspended solids results from surface run-off rather than leaching or direct input. A certain amount of suspended solids are sometimes found in drain flow from agricultural drainage schemes, but while these

Table 5.1 Possible soil conservation practices (after Stewart *et al.*, 1975).

Run-off control practice	*Effect on run-off*
No tillage planting in prior crop residues	Variable effect on run-off from substantial reduction to increases on soils subject to compaction.
Conservation tillage	Slight to substantial reduction.
Rotations including grass	Substantial reduction in grass years, slight to moderate reduction in crop years.
Rotations not including grass	None to slight reduction.
Winter cover crops	Slight increase to moderate reduction.
Improved soil fertility	Slight to substantial reduction, depending on existing fertility level.
Timing of field operations	Slight reduction
Contouring	Slight to moderate reduction.
Graded rows	Slight to moderate reduction.
Contour strip cropping	Moderate to substantial reduction.
Terraces	Slight increase to substantial reduction.
Grassed outlets	Slight reduction.
Drainage	Increase to substantial reduction.

may be significant for the phosphorus or pesticides they may carry, it is unlikely that sufficient soil will move via this route for the suspended solids themselves to be any problem.

Table 5.1 lists a number of possible soil conservation practices and evaluates their effectiveness at reducing run-off.

It can be seen that the effectiveness of many of the practices can be very variable. For example, in some circumstances sowing cover crops post harvest to avoid having bare ground exposed to rainfall in winter can result in moderate reductions in surface run-off. In other circumstances, for instance in northern latitudes where growing seasons are short, the cultivation necessary to establish the cover crop may result in more run-off occurring rather than less as the establishment of late-sown cover crops is often poor. As has been noted above in the case of BMPs to control nitrogen pollution, detailed knowledge of the local agricultural systems, the local climate and the local soils is necessary to select suitable techniques to reduce surface run-off.

Table 5.2 lists a number of soil erosion mechanisms and possible control methods. A range of control techniques is examined below.

5.8.1 BMPs to control suspended solid pollution

All the following can, in the correct circumstances, reduce soil erosion by water or reduce the transfer of eroded soil to water bodies. Not all are applicable in all circumstances.

Table 5.2 Erosion mechanisms and control

Mechanism	*Control*
Particle detachment by raindrops	Crop residues left on soil surface Crop cover Soil structural stability (increased organic matter content)
Particle movement by surface run-off (sheet erosion)	Improved soil structure and infiltration Improved land drainage Diversion terraces
Particle detachment and movement by running water (rill and gully formation)	Crop cover Grassed hollows Diversion terraces Check dams

Selection of crops protective of vulnerable soils

Some crops, such as grass and cereals, offer the soil far more protection from raindrop action than others, such as most row crops. Local knowledge is essential however as selection of suitable crops is not straightforward. For example, autumn-sown cereals, which in many areas are advocated over spring-sown cereals as offering better protection to soil, have proven to be markedly worse than spring-sown cereals in Scotland, probably because of the slow autumn growth in the northern latitude.

Early sowing of autumn-sown crops

If autumn-sown crops are well established before winter, increased soil protection results. Unfortunately more pesticides may also be required to protect the crop.

Growing green manure crops post harvest

In crops such as potatoes where the crop residue offers poor protection to the soil, growing a green manure crop may improve matters.

Conservation tillage practices

No till systems, minimal till systems. (Moldenhauer *et al.,* 1994; Conservation Technology Information Centre, 1996; Carter, 1998) The mouldboard plough has been the mainstay of European agriculture. However, one of its main benefits, weed control, can now often be done chemically. No till involves soils not being cultivated, with the next crop drilled directly into the residue of the previous crop. Minimal cultivation systems use shallow, non-inverting cultivators, which also leave crop residues near the surface. These crop residues directly protect soils from erosion and the high organic matter layer produced over time in the top few centimetres gives rise to well-structured stable soil which reduces rainfall run-off by increasing

Table 5.3 Mechanisms and processes involved in erosion control by conservation tillage (after Blevins *et al.*, 1998)

Conservation tillage and erosion processes			
Decrease erosivity		*Reduce erodibility*	
Decrease impact raindrop	Reduce run-off velocity	Enhance structure and aggregate stability	Enhance infiltration and decrease run-off

the infiltration capacity of the soil and reduces the detachment effect of raindrop impact. Lower erosion rates result. Pesticide use generally increases however.

Table 5.3 illustrates the basic ideas of conservation tillage methods.

Timing of cultivations to avoid pans and capping

Cultivating or otherwise trafficking soils in wet conditions gives rise to compact layers or pans, which reduce the infiltration capacity of soil and increase erosion.

Crop residue mulches

Leaving crop residues on the soil surface protects it from raindrop impact. Crop residues may harbour plant diseases and increase pesticide need, however.

Maintenance of adequate soil organic matter levels

Soil organic matter content is the key factor in maintaining good soil structure. Practices such as straw burning which destroy organic matter are generally undesirable.

Chemical soil stabilization

Certain chemicals, such as cellulose xanthate, have been used to stabilize soil structure to resist soil erosion, particularly wind erosion. However, they are expensive and the effects are short lived.

Contour strip cropping

Growing crops in alternating strips of erosion-prone and erosion-resistant crops can reduce total soil loss. Cultivating with machinery working along rather than up and down slopes also generally helps.

Land drainage and maintenance of existing drainage systems

Land drainage with piped drains laid across slopes reduces the risk of surface run-off of water and hence erosion in soils with slowly permeable subsoils such as many glacial soils. The technique has additional drainage benefits, but is very expensive.

Diversion terraces

Diversion terraces, i.e. regularly spaced channels running across slopes collecting and conveying water at low velocity to suitable discharge channels, usually grassed waterways, offers good erosion control at a price. They are widely practised in North America.

Grassing run-off carrying depressions in fields

Depressions in fields which run-off water erodes on a regular basis rarely give good crops and are often better put down to permanent grass.

Buffer strips

Buffer strips, that is linear areas of uncropped land normally, but not necessarily, adjacent to water courses may protect that water by trapping eroded soil before it enters the water. In cases where the run-off water is concentrated by the topography, the effectiveness is likely to be low.

Avoidance of over-grazing

Overgrazing damages grass swards (Sansom, 1999b) and reduces the protection provided to the underlying soil. A severe problem in dryland areas, its significance has often been overlooked in wetter upland regions.

Avoidance of poaching

Poaching, that is the damage to or destruction of grass swards in wet soils by animals' hoofs, can result in soil erosion in wet areas or areas with moisture retentive soils.

5.9 MICRO-ORGANISMS

Over recent years, there has been a growing realization of the possibly harmful effects of micro-organisms derived from livestock farming entering water (Mawdsley *et al.*,1995). A number of studies in the UK (Wyer *et al.,* 1997) have shown that the failure of compliance of bathing waters with EC micro-biological standards have not been due to sewage inputs, as has usually been assumed, but has been, on occasion, due to micro-organisms derived from livestock enterprises. That these studies have been done with seawater on beaches indicates the possible scale of the problem.

Most livestock manure receives little or no treatment before it is spread on land. Guidance on spreading techniques in codes of practice has largely focused on rates of spreading, conditions in which the manures are spread, and soil and topography on which they are spread. The aim of the guidance has been to avoid surface run-off washing the manure directly into any watercourse.

It should not be considered that it is only the spreading of stored manures and slurries which can cause problems. Grazing animals defecating directly in fields is also likely to give rise to water pollution.

The small size of some micro-organisms, particularly the viruses which are believed to cause most problems with water-borne disease, would suggest that transmission through soil to underlying field drains or permeable gravels is a distinct possibility.

5.9.1 BMPs to control micro-organism pollution

Spreading manures and slurries at appropriate rates

Spreading manures at too high rates increases the risk of pathogens being washed from the field to water.

Spreading manures and slurries in appropriate weather conditions

Spreading manures in conditions where soils are saturated with water or are frozen or snow covered increases the risk of surface run-off washing pathogens from the field to water.

Spreading manures and slurries on appropriate fields

Spreading manures on steeply sloping fields or fields with soils of low permeability increases the risk of pathogen transfer.

Injection of slurries

Equipment exists to inject liquid manures and slurries below the soil surface. The risk of run-off is greatly reduced. Higher rates of liquid can be safely injected than can be safely surface spread. Care must be taken however in fields with piped underdrainage.

Having an adequate area of land for stock carried

If farmers have an adequate area of land on which to spread manures and slurries, the need to spread at high rates is reduced.

Having an adequate storage capacity for manures and slurries

If farmers have an adequate storage capacity for manures and slurries, the need to apply in unsuitable weather conditions is reduced. Batch storage and distribution may reduce risks of pollution by active pathogens, by storing older material and distributing it without addition of further fresh material to that batch prior to application to fields (see next paragraph).

Farm treatment of manures and slurries to reduce pathogens: composting and digestion

Many stock farms have a population equivalent, in terms of pathogens produced, that is comparable with small to medium-sized towns. Sewage treatment technology such as composting or sludge digestion, to achieve high temperatures and kill pathogens, may increasingly be seen to be desirable in sensitive areas.

Treatment for unavoidable, residual contamination of farmyard run-off

Even after rigorous separation of dirty water (slurry, dairy washings), water from relatively uncontaminated yard areas and roads could still be a significant source of pollution if all farms in a sensitive catchment are considered together. A retention pond or stormwater wetland for steading run-off should provide a best practice level of protection (see design details in Chapter 3).

Avoiding stock having direct access to watercourses

The provision of water troughs, rather than allowing stock to drink directly from watercourses, will reduce pollution risks.

Buffer strips

By keeping stock away from water, buffer strips have a role to play here too.

5.10 OVERVIEW

A wide range of possible BMPs have been cited above as possible solutions to a range of potential problems. Table 5.4 summarizes the possible role of some of the more important ones.

5.11 TARGETING BMPS ON CRITICAL AREAS

Often, most of the diffuse pollution from a farm can be shown to arise from only a small proportion of the area of the farm. Much of the land will frequently be too remote from watercourses to cause much of a problem. The costs of introducing BMPS can often be reduced greatly by targeting them to the critical areas of the farm where pollution of water is likely to occur. For example, low rates of soil erosion in areas of the farm remote from watercourses where eroded soil is likely to be deposited in fields rather than into water may be of little consequence. Efforts should be concentrated on areas where transfer of pollutants from the land to water is likely to occur.

Table 5.4 Measures that offer general benefits and pollutants they may be best suited to controlling.

	P	NO_3	*Faecal paths*	NH_4	*BOD*	*SS*	*Pesticides*
Waste management plans	+	+	+	+	+	(+)	–
Nutrient budgets	+	+	–	(+)	–	–	–
Buffer strips	(+)	–	+	?	?	(+)	+
Contour ploughing	+	–	?	?	–	+	?
Farm steading ponds or reedbeds	+	+	+	+	+	+	+
Composting	?	+	+	–	+	–	?
Slurry digestion	?	+	+	?	?	–	?
Reducing livestock density	+	+	+	+	+	+	?

5.12 INTERACTION BETWEEN BMPS

Often the introduction of one BMP to a farming system may reduce one problem but may, at the same time, increase others. For example, no till and reduced tillage systems of farming greatly reduce soil erosion and its associated pollution. However, the need for pesticides, both herbicides to control the weeds that are no longer being killed by ploughing and pesticides to control plant diseases which can be harboured in the plant residues now left at the surface, will usually be increased. If the water body that was being impacted by diffuse pollution, was being damaged by pesticides and if soil erosion was not an important transfer mechanism from land to water, introducing no till or minimum till systems would be foolish. The problem being addressed is likely to get worse, not better.

In some of the brief descriptions of BMPs given above, it has been noted on occasion that certain techniques may not be effective in all geographical areas. For example, sowing green manures post harvest to reduce nitrogen leaching works well in most areas, but has been shown often to be worse than useless in the cold northern climate of Scotland. The green manure simply does not grow well enough to take up enough nitrogen to be useful.

In all cases, a considerable degree of local knowledge will be necessary before appropriate BMPs can be selected for a given farming system in a particular area. Much of that local knowledge will be held by the farmers themselves. Their active involvement not just in the implementation of a pollution reduction policy but also in its planning is essential.

5.13 PERSUADING FARMERS TO ADOPT BMPS

The key issue for BMPs is, of course, the extent to which farmers can be persuaded to adopt them. In this respect, BMPs fall into two groups, those which it is economically beneficial for the farmer to adopt and those which will cost the farmer money.

The first group, BMPs that show an on farm cost benefit, should be easier to persuade farmers to implement. An example is using animal manures in a cost-effective way. Many farmers, particularly dairy farmers, have come to regard animal manure, and particularly cattle slurry, as a waste product to be disposed of. In fact, slurry has a nutrient value of about £3.00 per cubic metre (at 2004 prices) and in many cases can totally replace artificial fertilizer within the farm. However, large numbers of farmers continue to ignore the fertilizer value of animal manures and to add the same rates of chemical fertilizer as they would on fields where animal manures are not applied. Another example is the continued prevalence of arable farmers adding phosphates to soil at rates far in excess of off-takes of the nutrient in the harvested crop. Such excess inputs were agriculturally cost effective in the past when many soils were inherently deficient in phosphorus. This, however, is rarely the case today for arable land in developed countries and farmers could very frequently greatly reduce their inputs of chemical phosphates with no yield penalty. Persuading farmers implement BMPs in this group ought to be relatively easy as it can be clearly shown to be in the farmers' best financial interest to do so.

The other group of BMPs, those where implementation will cost the farmer money, are more problematic. Examples would include reducing stock numbers to a level where the manures produced by the stock on the farm can be disposed of on the land available to the farm without causing a continuing rise in the nutrient status of the soil. In intensive livestock enterprises, inputs of nutrients in the form of animal feed may far exceed the off-takes from the farm system in the form of animals sold off the farm. The extreme of such situations are probably American 'feedlot' farms where very little of the nutrition of the stock is derived from crops grown on the farm. The only real sustainable solution is to reduce stock numbers. Practices being introduced in America where resulting manures are carried long distances (sometimes hundreds of miles) to disposal areas are clearly undesirable.

Another example of a best practice, which will generally cost money on the farm, is erosion control. While this is undoubtedly desirable to reduce the transfer of pollutants (silt, adsorbed phosphorus, adsorbed pesticides) from the land to water, it will frequently cost the farmer money. Soil erosion, which is not sufficiently severe to reduce the productivity of the land, at least in a time scale of centuries, may still have serious consequences for the quality of receiving water. Therefore, such techniques as diversion terraces are unlikely to be adopted without some incentive as they cost money to construct and use up land which could otherwise be producing crops. (Frost *et al.*, 1990)

Financial incentives or coercion may be necessary to persuade farmers to adopt practices which cost them money, even if those practices show a clear overall cost benefit when environmental costs are taken into consideration. Some environmentally minded individual farmers may adopt them but many will resist. This is particularly so at a time when farming income is under pressure. To quote one Scottish farmer, 'It is hard to be green when you're in the red'.

Coercion is difficult to implement and often counter-productive. As has been discussed, it is hard to prove a link between the undesirable effects of diffuse pollution and any individual farm business. Nor is forcing farmers to adopt practices which reduce farm incomes a good way to ensure that the practices are enthusiastically and effectively applied. Policing implementation of many BMPs is virtually impossible.

The best way to encourage adoption of this group of practices is to make it financially advantageous for farmers to do so. This can usually only be done by targeting state support to agriculture to achieve these objectives. That does not necessarily mean extra payments; this issue is considered, with examples, in Chapters 7 and 8.

APPENDIX 5.1 SAMPLE WORKED NUTRIENT BALANCES

The most useful nutrient balance from the environmental standpoint is a phosphorus balance. Potassium balances can also be very useful agronomically but potash rarely presents any environmental threat. Nitrogen balances can be calculated but generally contain too many large unknowns such as legume- or clover-fixed nitrogen and nitrogen released from soil organic matter breakdown to be very useful.

Whole farm balance, phosphorus, for a dairy farm with a poorly designed fertilizer policy

Inputs (expressed in terms of P, not P_2O_5)

Fertilizer

14 hectares silage land at 31 kg/ha	434 kg
45 hectares grazed grass at 6 kg/ha	270 kg
Fertilizer P total	**704 kg**

Bought-in feed

240 tonnes draff at 0.14% P	336 kg
80 tonnes beet pulp at 0.14% P	12 kg
40 tonnes maize gluten at 0.8% P	320 kg

60 tonnes straw at 0.07% P	42 kg
88 tonnes 18% cake at 0.6% P	528 kg
36 tonnes 16% calf feed at 0.6% P	216 kg
Feed P total	**1454 kg**
Total bought-in phosphorus (P)	**2158 kg**

Off-takes

500,000 litres milk at 0.001 kg P/l	500 kg
52 1-year-old cattle at 3 kg P each	156 kg
15 cast cow at 4 kg P each	60 kg
Total off-takes of phosphorus (P)	**716 kg**
Net whole farm balance	**+1442 kg P**

The whole farm inputs of phosphorus therefore exceed the whole farm offtakes by 1542 kg/annum. This averages out at 26 kg phosphorus (59 kg P_2O_5) per hectare of improved grass per annum.

Whole farm balance, phosphorus, for a mixed farm with a well-designed fertilizer policy

Inputs (expressed in terms of P, not P_2O_5)

Fertilizer

17 hectares 1st wheat at 26 kg/ha	442 kg
35 hectares other wheat at 26 kg/ha	910 kg
17 hectares oats at 26 kg/ha	442 kg
34 hectares barley at 26 kg/ha	884 kg
24 hectares silage at 0 kg/ha	0 kg
63 hectares grazed rotational grass at 0 kg/ha	0 kg
70 hectares permanent grazed grass at 0 kg/ha	0 kg
Fertilizer P total	**2678 kg**

Bought-in feed

30 tonnes 18% protein ewe concentrate at 0.6% P	180 kg
32 tonnes 14% protein lamb pellet at 0.6% P	192 kg
14 tonnes soya at 0.5% P	70 kg
Feed P total	**442kg**

Bought-in stock	
100 ewe lambs at 0.3 kg/lamb	30 kg
5 tups at 0.5 kg/tup	2.5 kg
Bought in stock P total	**33 kg**
Total bought-in phosphorus (P)	**3153 kg**

Off-takes

429 tonnes wheat at 3.4 kg/tonne	1459 kg
117 tonnes oats at 3.4 kg/tonne	398 kg
70 tonnes straw at 0.6 kg/tonne	42 kg
1650 lambs at 0.28 kg/lamb	462 kg
20 heifers at 3.9 kg/beast	78 kg
30 bullocks at 4.8 kg/beast	144 kg
150 cast ewes at 0.5 kg/ewe	75 kg
7 cast cows at 4 kg/beast	28 kg
Total off-takes of phosphorus (P)	**2686 kg**
Net whole farm balance	**+467 kg P**

The whole farm inputs of phosphorus therefore exceed the whole farm off-akes by 467 kg/annum. This averages out at 1.7 kg phosphorus (4 kg P_2O_5) per hectare of arable land and improved grassland land per annum. This is an excellent figure as close to true balance as is practicable.

APPENDIX 5.2 SAMPLE FARM WASTE MANAGEMENT PLAN

A full farm waste management plan is a substantial document covering the following:

- Description of systems in place for collection, storage, transport and land application of slurry, farmyard manure and contaminated water;
- Detailed information about each system, including storage capacity, how it is operated and how these operations are recorded and how the system is monitored and maintained;
- Audit of organic manures and contaminated water produced;
- Details about the available spreading area on the unit with field plans and identified risks (classified high, medium or low) associated with each of the fields concerned;

- Details about days available for spreading taking account of average monthly rainfall, dry days and frost days;
- Calculations to demonstrate that the systems in place or planned are adequate to deal with the quantities of organic wastes produced. (After Henderson, 2002, Scottish Environment and Rural Affairs Department information note).

The plan is an active document meant to be consulted and, if necessary, updated regularly. What follows only covers certain areas of such a plan.

Part of a farm waste management plan for a small organic dairy farm with 42 milking cows

Volumes and nutrient contents of manures and slurries

Slurry, 29 week winter period

42 dairy cows contribute slurry to the lagoon, being housed for 29 weeks. Volumes and nutrient contents are tabulated below.

42 dairy cows housed for 29 weeks:

Slurry volume (m^3)	Nitrogen (kg)	Phosphate P_2O_5 (kg)	Potash K_2O (kg)
450	2249	890	2249

Wash water once daily for 29 weeks. (allow 18 litres per animal per day)
Volume (m^3) = 153

Rainwater entering slurry lagoon in 29 week winter period less evaporation.
Volume (m^3) = 110

The winter (29 weeks) slurry totals will therefore be a volume of 713 cubic metres containing 2249 kg of nitrogen, 890 kg of phosphate and 2249 kg of potash.

Slurry, 23 week summer period

In this period, the cows are brought into the byres only for milking. Slurry will be limited the dung and urine deposited during milking, to wash water and to rain water. Volumes and nutrient contents are estimated below.

Slurry volume (m^3)	Nitrogen (kg)	Phosphate P_2O_5 (kg)	Potash K_2O (kg)
42	210	83	210

Wash water twice daily for 23 weeks. (allow 25 litres per animal per day)
Volume (m^3) = 201

Rainwater entering slurry lagoon in 23 week summer period less evaporation.
Volume (m^3) = 29

The summer (23 weeks) slurry totals will therefore be a volume of 272 cubic metres containing 210 kg of nitrogen, 83 kg of phosphate and 210 kg of potash.

Farm yard manure

Farm yard manure is produced by the 8 dairy cows housed in the small byre for 29 weeks and by the young stock housed in the cattle court for about 19 weeks. The young stock comprise about 25 animals below twelve months old and about 30 animals aged from twelve to thirty months old. 80 tonnes of straw are used in total.

Together these will produce about 350 tonnes of farm yard manure. This will contain when fresh about 2100 kg of nitrogen, 1225 kg of phosphate (P_2O_5) and 2800 kg of potash (K_2O).

Land available for spreading manures and slurry

The map in Figure 5.5 divides the land into three categories. That marked "+ + +" is assessed in terms of soil type, topography, proximity to open water etc. as being of lower risk for spreading manures and slurries. There are 44.3 hectares of such land.

That marked "L L L" is assessed as being of higher risk for spreading of manures and slurries, in all cases here because of the nature of the soil. There are 10.1 hectares of such land. Its use should be restricted to the summer months.

There was no "Medium" risk land.

That left blank is assessed as unsuited for the spreading of manures and slurries. This is for a variety of reasons, ranging from proximity (within 10 metres) to water courses or open water, wetlands, roads, buildings and steading etc. No manures or slurry should be spread on this land.

Rates of nutrient application

Because this is an organic farm, no chemical fertiliser is brought in. All the nutrients supplied to the grass come from the manures and slurries or from fixed atmospheric nitrogen from the clover in the sward.

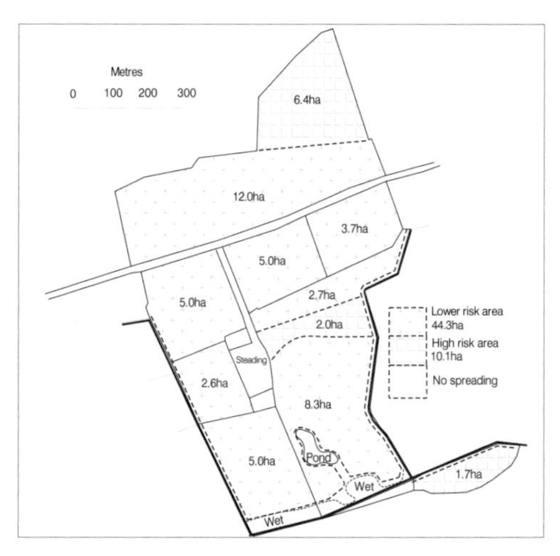

Figure 5.5 Sample map

The slurry is all spread onto the silage ground. 21.3 hectares of fields 1,2,4,7 and 9 is assessed as of lower risk. The slurry will supply a maximum of 115 kg/ha N, 46 kg/ha P_2O_5 and 115 kg/ha K_2O to this land. In practice the nitrogen actually applied will be about 72 kg/ha because of losses in storage and at application and not all that will be available to the grass because of losses from the soil.

The farm yard manure is spread over all the available land (54.4 hectares). It will supply a maximum of 39 kg/ha N, 23 kg/ha P_2O_5 and 51 kg/ha K_2O to this land. In practice the nitrogen actually applied will be significantly less (about 25 kg/ha) because of losses in transporting it and in storage and not all that will be available to the grass because of losses from the soil.

Total nutrient inputs from manures and slurries are therefore likely to be about 97 kg/ha N, 69 kg/ha P_2O_5 and 166 kg/ha K_2O to the silage land and 39 kg/ha N, 23 kg/ha P_2O_5 and 51 kg/ha K_2O to all other land assessed as suitable for spreading.

None of these figures give any cause for environmental concern and simply illustrate the dependence of organic systems on atmospheric fixed nitrogen.

Volumes of manure and slurry application

The farm has about 350 tonnes of farm yard manure to spread on 54.4 hectares, a mean rate of 6.4 tonnes per hectare. This is a low rate and gives no environmental cause for concern.

There are a total of 713 cubic metres of slurry to spread in winter (October to May) and 272 cubic metres of slurry (mainly dirty water) to spread in summer

(May to October). The winter slurry is spread on 21.3 hectares of suitable silage ground because the silage needs the nutrients in the slurry. The summer dilute slurry could go anywhere suitable as it contains little by way of nutrients.

The winter slurry amounts to 33 cubic metres per hectare of suitable silage land. The maximum rate at which it is advisable to surface spread slurry is 50 cubic metres per hectare per dressing. There is no reason why any of the suitable silage land should have more than one dressing therefore.

Timing of manure and slurry applications

The farm yard manure is removed from the courts to the field midden after the cattle go out in summer. The field midden should be made as deep and compact as possible to minimise nitrogen losses in storage. The manure is spread throughout the farm. Spring is the best time to maximise the efficiency of nitrogen use.

The winter slurry starts to accumulate in the lagoon from mid October. The lagoon is likely to be filling up at a rate of about 25 cubic metres per week so must be emptied by the end of December. It will require to be emptied again be early March and again by May. The first application is rather earlier than desirable from a standpoint of utilisation of nitrogen but gives no particular environmental concern. The silage land on which the slurry is to be spread is naturally imperfectly drained but has a relatively light (sandy loam) texture in the topsoil and upper subsoil. In most years, suitable conditions for spreading with a vacuum tanker will occur in the winter without spreading on saturated or frozen land. As a back-up for particularly wet years, a contractor using umbilical cord equipment is available. Such equipment can work without undue risk of damage to soil or to the wider environment in markedly wetter conditions than conventional spreaders. It is considered therefore that the existing slurry storage volume is adequate.

The spreading of the summer dilute slurry presents no particular problem.

Cattle poaching; soil damage and polluting runoff to watercourse avoided by locating feeders well away from drainage pathways for pollution (Scotland).

Fencing off access by livestock (Scotland).

Buffer strip alongside arable land, Wisconsin.

Grassed runoff pathways through fields to minimize erosion in the USA. (USEPA).

6

Rural best practice experience

6.1 ADVANTAGES AND DRAWBACKS OF A BMP APPROACH

The use of a BMP approach has a number of advantages. First, it is unnecessary to prove a sometimes complex and poorly understood link between the undesirable actions of the farmer and the pollution of the water. Such links are far harder and more expensive to prove (to the satisfaction of a sceptical court) than for classic point source pollution where the pollutant emerges from a pipe and has an evident impact on water quality over and above other influences.

Second, the approach itself directly presents the farmer with a course of action, which will lead to the solution to the environmental problem. The farmer is not being asked to do something (such as to reduce the level of phosphorus input to a receiving lake) which has little obvious direct connection to normal farming operations. Rather, he is being asked to carry out farming operations in a specific way (such as adopting particular cultivation practices in specified fields which will minimize the rate of soil loss from these fields and the rate at which soil-derived phosphorus enters the lake).

Third, the approach makes it easier for catchment specific targets to be set. Since the criteria for success or failure of a programme is assessed at the catchment level

 Diffuse Pollution by N.S. Campbell, B.J. D'Arcy, C.A. Frost, V. Novotny and A.L. Samson. ISBN: 1 900222 53 1. Published by IWA Publishing, London UK.

by assessing whether or not the perceived environmental problem has been solved, recommendations and programmes can be made catchment-specific with less of a risk of farmers within the catchment feeling that they are being unfairly disadvantaged when compared to other farmers outside the catchment.

Fourth, the approach is open-ended and encourages innovation. Because a whole range of BMPs may offer a potential solution to a given problem, the farmer, his advisors and agricultural researchers are encouraged to test a wide range of possible strategies and combinations of strategies.

The approach is, however, not without drawbacks. One major challenge is to ensure that the proposed measures will actually achieve the desired result, and to be clear that there is a sufficiently broad understanding as to what sort of benefits constitute the desired result. For example, buffer strips have been widely promoted as a landscape measures that have a role to play in controlling movement of diffuse source pollutants into watercourses, together with creation of complementary habitat and potential value as wildlife corridors. It would be a mistake to assume that such practices alone would be the best means to control a particular problem pollutant in local circumstances, for example to reduce the transfer of phosphorus from land to water. Even worse, cynics may seize on such measures and assess them for attributes for which they are clearly unsuited, and not advocated as best practice controls (e.g. controlling a highly water soluble contaminant such as nitrate from field drains; incredibly, research funds have been used to carry out such investigations!)

There is a risk with a BMP approach that the target of the regulators becomes simply the widespread adoption of BMPs. A target of so many miles of buffer strip installed and a scientifically devised nutrient budget and farm waste management plan for every farm are perfectly acceptable components of a catchment plan to address diffuse pollution. But it should not be forgotten that the bottom line, the ultimate goal in terms of water quality objectives, is the improvement of the water resource.

A related problem lies in the relatively long timescale of some of the expected benefits. Some BMP measures are likely to take a number of years or even decades to bear fruit. In such cases, a considerable amount of research is necessary to ensure that the proposed programme will actually produce the desired result, as the success or otherwise cannot be directly assessed for a number of years.

A number of case studies, which illustrate some of the above advantages and drawbacks in practice, are now examined..

6.2 LOCH LEVEN CATCHMENT, SCOTLAND

This lowland lake in central Scotland extends to some 13.3 km^2. It has a catchment of about 145 km^2. Eighty per cent of the catchment is agricultural land, eleven per cent woodland and two per cent urban areas, the balance being made up of various other land uses. The agriculture ranges from extensive grazing of sheep on the hills to intensive production of vegetables and potatoes on the better land closer to the lake.

The lake is naturally eutrophic, but the influence of man in the catchment led to a sharp drop in water quality over the twentieth century. The most obvious problem is a great increase in the algal growth in the lake in summer. As the lake is an important sport fishery for brown trout and is also of great importance to migratory and breeding wildfowl, studies have been undertaken to determine the causes of the problem.

Initial studies were undertaken to determine what nutrient (or other factor) limited algal growth in the lake.

The concept of a *limiting nutrient* needs a little further explanation. In order for algae to grow, the full range of normal plant nutrients including nitrogen, phosphorus, potassium, sulphur and a range of trace elements must be present. In normal circumstances, one of these nutrients (or some other factor such as light) will be in shortest supply and will *limit* the algal growth. In order to reduce algal growth, it is necessary to reduce the input of the limiting nutrient. For example, if phosphorus is the limiting nutrient, reducing somewhat the quantity of nitrogen entering the lake will have no effect on algal growth. Only when nitrogen inputs have been reduced to such an extent that it becomes the limiting nutrient (rather than phosphorus) will algal growth be reduced. However, reducing the input of the limiting nutrient, even by a relatively small amount, has an immediate impact on the algae.

Phosphorus was shown to be the limiting nutrient and further studies were undertaken to determine the sources of that phosphorus. As is frequently the case, multiple sources were identified. These are set out in Table 6.1.

Initial action was targeted at the point sources. A succession of changes in processing techniques of a woollen mill reduced the industrial input from 18 kg/day to less than 1 kg/day by 1988 (D'Arcy, 1991). Upgrading works in the local sewage treatment works to strip phosphorus reduced inputs from this source to about 3 kg/day, leaving diffuse pollution by far the greatest remaining source (Frost, 1996).

In 1995, a Catchment Management Project was initiated to attempt to find ways of reducing the phosphorus inputs further. This was a collaborative effort involving

Table 6.1 Sources of phosphorus entering Loch Leven in 1985

	Amount in kg per day
Industrial	18
Sewage treatment works	13
Diffuse sources, mainly agriculture	26

representatives of a range of relevant organizations including the government conservation agency (Scottish Natural Heritage), the government regulatory agency (Scottish Environment Protection Agency), the local authority planning department, the Scottish Agricultural College and representatives of interested parties including the National Farmers' Union and local farmers.

Work was carried forward by a number of working groups. These included the following:

- **a water quality group** charged with reviewing the water quality standards for the lake and with determining the required reduction in phosphorus inputs,
- **a river management group** charged with improving the habitat of the feeder streams to the lake by such measures as grant-aided marginal tree planting,
- **a planning and development group** charged with producing planning guidelines aimed at allowing further development of housing and industry in such a way that further phosphorus loading to the lake from these sources was prevented and
- **an agriculture and forestry group** charged determining the nature of the diffuse agricultural pollution and implementing measures to reduce it.

The project reported its finding in 1999 (Flint, 1999). Soil erosion at relatively low rates (1 to 2 tonnes per hectare per annum) was implicated as a major source of phosphorus input particularly from the heavily fertilized, intensively cropped vegetable and potato land. A range of actions were proposed to combat this soil loss to the lake.

All farmers within the catchment were sent a leaflet setting out a range of simple measures to reduce soil loss. These included small changes in current methods and timings of cultivations aimed at protecting soils at vulnerable periods. However, in the absence of any financial incentive, uptake has been poor. The low rates of erosion experienced in this catchment pose no threat to the viability of agriculture, even on a time scale of centuries. Almost all erosion control measures have a cost to the farmer, either in terms of increased costs or reduced yields and, without some further incentive, uptake is likely to remain poor.

As part of the Loch Leven project, farmers within the one intensively managed sub-catchment (the Greens Burn) were encouraged to leave a buffer strip of varying width between the lake feeder stream and the cropped land, driven by three aims:

- habitat enhancement (the Farming and Wildlife Advisory Group was a key partner in that initial project)
- prevention of pesticide overspraying in that intensive vegetable growing catchment (biological monitoring of water quality had indicated a generally poor fauna, subject to intermittent catastrophic plunges in quality that were thought to be related to insecticide applications
- contribution to control of soil erosion.

Uptake here was better as a consequence of one-to-one persuasion and explanation of need to the farmers. Some measures were established utilizing a government grant scheme that was available to meet much or all of the cost. Others were established as part of the requirement to put some land into non-productive set-aside, whilst others were provided at the farmer's own initiative. Some 14 km of buffer strip were left and planted to grass. This represents all of both banks of the stream where sediment transfer was likely to occur. Subsequently on another tributary, where mixed farming predominated, buffer zones were also established as part of the Loch Leven project, by fencing off access for livestock.

In-field measures were also sought, across the lake catchment, in order to tackle problems of leaching of phosphorus into drains and water courses. A major programme of soil analysis was carried out with soil samples being drawn from 141 randomly distributed fields. On those farms where soil phosphorus levels were shown to be higher than necessary for optimal crop yields, farmers were visited and a nutrient balance sheet programme of fertilizer needs drawn up. In such cases these programmes usually result in savings for the farmer in the cost of bought-in fertilizer with no consequent loss in yield. They are therefore usually adopted.

Other measures were also taken to address agricultural sources. All farms in the catchment were visited; advice was given to farmers to relocate field middens away from water margins, to prevent direct leaching of phosphorus-rich drainage; and at a few livestock steadings various measures were needed to address sources of drainage that should not have been reaching the watercourses.

The results in terms of water quality of these efforts require long-term catchment assessments.

Preliminary results for the Greens Burn tributary of Loch Leven suggest that the biological quality has improved in that the periodic catastrophic plunges in invertebrate field scores have stopped since the buffer strips were provided. It is too early to assess achievement, or otherwise, of the other two objectives for the Greens Burn buffer strips.

Technical difficulties have so far prevented sufficient data collection for the demonstration of the relative effectiveness of the rural BMPs to protect and restore water quality. The Loch Leven case study is, however, an important example because the combined impact of all the remedial actions taken (major point as well as diffuse sources) has become evident in the water quality of the loch.

Figure 6.1 shows the improvement in water quality as indicated by the maximum water depth in which macrophytes will grow since the project began. While not back to the levels of the early twentieth century, the improvement since the 1980s is marked. The restoration of macrophyte cover (inhibited by lack of light penetration associated with planktonic algal growth) was the primary ecological objective for the Loch Leven management plan.

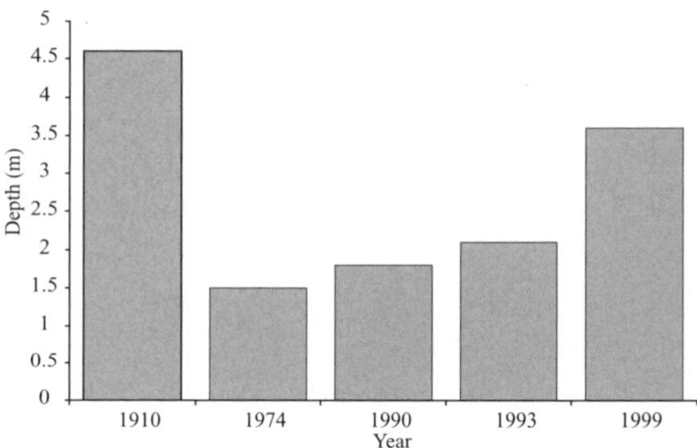

Figure 6.1 maximum depth at which macrophytes are recorded in Loch Leven, Scotland (Lynda May, unpublished note, May 2003, Centre for Ecology and Hydrology, Penicuik).

Text Box 6.1 Buffer strips

Buffer strips are strips of uncropped land, frequently planted with grass or grass/wildflower mixes, left along the edge of rivers or lakes. The aim is to protect the water body from pollution, and create wildlife corridors and useful additional wildlife habitat on the farm.

If they are to fulfil the pollution prevention aim, they must be properly designed to meet the needs of the individual problem. In particular, the nature of the pollution problem must be assessed and the buffer strips tailored to treat that particular problem.

All buffer strips help to protect water against poor farming practice such as allowing fertilizer spreading or spray drift to directly enter water. Even buffers as narrow as two metres can be effective in some circumstances, although at least 15 metres is generally needed for sediment control (Davies and Christal, 1996).

Most ordinary buffer strips will do little to reduce nitrogen pollution as this nutrient mostly moves in solution in soil water. Some nitrogen will be taken up by the growing plants and held in organic form, but the quantities taken up are generally not great. In order for buffer strips to tackle nitrogen problems effectively, the soil in the buffer strip must be wet. This can sometimes be achieved by constructing river margin wetlands, perhaps by stopping field drains short of the river to discharge into the wetland. In wet soils, with adequate supplies of organic matter coming from the wetland plants, denitrification (the reduction of nitrate nitrogen to gases such as nitrogen or nitrous oxide) will occur. The nitrate, which would otherwise have polluted the river, is lost to the atmosphere.

River margin wetlands can only be constructed successfully where the topography and soils are favourable. For example, on very flat sites the loss of hydraulic head involved in discharging field drains to the wetland may adversely affect the land drainage further away from the river. However, where they can be successfully built, they offer a good method of reducing nitrogen pollution.

Phosphorus may move in particulate form adsorbed onto moving soil particles, may be in solution, or may move in colloidal form. Buffer strips may be effective in removing phosphorus moving in particulate form as the eroded soil may be trapped in the buffer strip and accumulate, with its adsorbed phosphorus, there. This is discussed further below.

Phosphorus moving in solution or in colloidal form will frequently be in field drains. Unless these drains are stopped short of the water course the buffer strip will have no effect. On many sites it is difficult to effectively stop drains short, without affecting the whole drainage system.

Dissolved and colloidal phosphorus moving through the soil matrix itself may be adsorbed if it passes through soil of low phosphorus status. If riverside soils have never been fertilized this may be the case, but frequently buffer strips are of necessity placed on soils which have the same chemical fertility as the rest of the field. If the rest of the field is of high phosphate status, allowing phosphorus to leach, the buffer soils are unlikely to be able to adsorb that phosphorus. If the water flow is mostly confined to the topsoil because of impermeable subsoil, one technique which might be applied is deeply ploughing the buffer strip to mix infertile subsoil through the topsoil to increase the phosphorus adsorption potential.

The main action of buffer strips on **pesticides** will be to reduce the risk of spray drift directly to the watercourse. Provided farming practice is good and pesticides are not spread in unfavourable conditions where drift is likely, narrow buffers of two metres may be effective. Pesticides moving adsorbed onto soil particles may also be trapped (see below).

Suspended solids, resulting from soil erosion, may in some circumstances be trapped in densely vegetated buffer strips. An effective buffer would be one with dense low vegetation, low gradients across the buffer from field to river and situated where the field topography did not concentrate sediment carrying run-off into defined channels. No buffer will stop sediment carried in water running several centimetres deep. Wider buffers are more likely to be effective than narrow ones, but in some circumstances even two metres can help.

Buffers to trap suspended solids often will be more effective if placed in hollows within the field where water frequently is known to flow rather than directly next to the watercourse to be protected. Care must be taken with both grass strip positioning and with subsequent cultivations to avoid running water forming rills along the edge of the grassed strip.

Micro-organism pollution can be reduced using buffer strips by keeping animals and dung spreaders away from watercourses reducing the risk of contaminated material being washed to the river.

Text Box 6.2 Nutrient balance sheets

Nutrient balance sheets are a powerful tool to aid arable farmers in planning fertilizer use. Balances are calculated on an individual field basis. The soil is regarded as a 'bank account' for nutrients, the inputs being applied chemical fertilizer and the nutrients contained in animal manures and slurries. The off-takes are the nutrients contained in the fraction of the crop which is removed from the field at harvest. If the inputs and off-takes are equal, the nutrient status of the soil will remain constant. A balance sheet approach is only useful for the relatively non-mobile nutrients, such as phosphorus and potassium. The rates of loss of these nutrients by leaching are, in most soils, negligible for agricultural purposes (although not so in terms of their impact on the environment where agriculturally minor losses may be very damaging). Nitrogen fertilizer is subject to large leaching losses and a balance sheet approach will not work.

Initially the farmer must have the soil in his fields analysed. If it has agriculturally optimal levels of phosphorus and potassium, he will from then on only apply sufficient of these nutrients, including the total quantities present in any animal manures or slurries applied, to meet the off-take in the harvested fraction of the crop. Straw etc. that is left in the field does not count. If the soil levels were less than those which are agriculturally optimal, he is justified in applying extra nutrients to build up the soil. If the soil levels are above what is agriculturally optimal, he may cut back on nutrient inputs and allow soil levels to fall.

In practice this approach frequently results in farmers applying less phosphorus than they previously would have, with beneficial results both to the farmers pocket and the environment. The approach also should avoid soil nutrient levels building up to undesirably high levels where leaching becomes a significant environmental problem.

Farmers can introduce more sophistication into the system by applying all or most of their phosphorus and potassium to those crops in the rotation likely to respond (such as potatoes) and apply little or none to those unlikely to respond (such as cereals). Provided the nutrient balance is maintained over the rotation, no yield loss will result.

One problem may be aiming for agriculturally optimal nutrient levels. These are determined by traditional crop trials where yield is the main factor in decision making. The level will vary with soil type. However, there may also be a soil nutrient level which ought not to be exceeded for environmental reasons. This will be the nutrient level at which nutrient losses from the soil exceed the capacity of the particular catchment to absorb those losses without unacceptable environmental damage (see Figure 5.3). If the environmental maximum level is above the agriculturally optimal level, there is no problem. If it is less than the agriculturally optimal level, conflict is likely. Unfortunately no reliable accepted method yet exists to determine this environmentally maximum nutrient level. It is the subject of much research interest.

Nutrient balance sheets are generally applied by farmers on arable fields or those where grass is grown for hay or silage. Account can be taken of applied animal manures. It is more difficult to use in grazed fields as the additional complications make the system unwieldy. A variant can be used for stock farmers using whole farm nutrient budgets. Here inputs will be bought in fertilizers and manures and bought-in feed. Off-takes will be the nutrients in milk or animals sold off the farm plus the nutrients in any crop sold off the farm. While less useful than field by field balances, potential problems can still be highlighted and addressed before harm is done.

6.3 SNY MAGILL CREEK, IOWA (USEPA, 1997b)

This trout steam in Iowa has been identified as having impaired water quality due to diffuse sources, mainly of agricultural origin. Pollutants include sediment, animal waste, nutrients and pesticides. It has a catchment of about 90 km^2, comprising row crops, pasture and forest. It includes about 140 grain, dairy, beef and pig producers. The catchment is characterized by gently sloping uplands falling to steep slopes.

Adoption of soil conservation BMPs since 1991 (mainly contour terraces, water and sediment control basins, contour strip cropping, and conservation tillage) is estimated to have reduced sediment delivery to the creek by over 40 per cent.

Stream bank erosion is thought to be a major source of sediment. To combat this, a programme of bank stabilization has been implemented using such techniques as willow posts and rock riprap.

Landowners have been encouraged to adopt animal waste management plans providing an economic benefit to the farmer and a reduction of nitrogen and phosphorus inputs to the environment. About 30 plans had been developed by 1997.

Integrated crop management methods have been adopted to help producers balance the nutrient and pesticide application with plant and soil needs. Using such methods, nutrient inputs to the catchment have been reduced by 18 tonnes of nitrogen, 15 tonnes of phosphate and there has been a reduction in use of corn rootworm insecticide. These techniques have resulted in an average saving to the farmer of about $35 per hectare.

Text Box 6.3 Farm waste management plans

Farm waste management plans are a way of aiding farmers to make decisions concerning the management of potentially polluting farm wastes, particularly animal manure and slurries.

An inventory of all the manure and slurry likely to be produced is drawn up alongside the available land for disposal. Rates of manure and slurry spreading are set having regard to such things as the likely uptake of nutrients by the crop and the current levels of the nutrients in the soil.

The timing of applications is set by such factors as soil type, climate, topography and crop nutrient demand, to avoid the risk of nutrients being washed from the land by surface run-off or leaching.

The final plan will indicate where, when, and at what rates manure and slurries can be safely spread. It will also indicate if additional storage capacity for manure and slurry is needed to ensure that it can be safely applied in suitable weather and soil conditions, and at a time when the nutrients are unlikely to be lost by leaching before the crop can take them up.

Farm waste management cannot however solve the environmental problems caused by farmers who simply have too much stock for the land available for manure and slurry disposal.

6.4 FRENCH CREEK, CALIFORNIA, USA

(Watershed Information Network News, 1998.) This catchment of 82 km^2, which lies in northern California, had a problem of excessive quantities of sediment, mainly sand, entering the creek. A mixed group of ranchers, commercial timber companies, a property owners' association and other interested parties was established to address the problem. It was determined that over 60 per cent of the sediment derived from forest roads, mainly constructed on the highly erodable granitic soils of the area.

A total of $500,000 was spent (20 per cent grant-funded) on road improvements resulting in the recontouring and rock surfacing of 60 km of road, the closing of 7 km of road (more in wet periods), the planting of 20,000 trees on cut-and-fill slopes, the placing of rock breast walls and rock mulch on large roadcuts and the rock surfacing of 7 km of private driveways.

In only three years, this work reduced the sediment entering French Creek by 75 per cent with the benefit that Coho salmon and steelheads returned to the water again. In-stream evaluation of fish numbers and sediment levels continues.

This project involved voluntary action by the main polluters without the need for regulatory coercion. Each partner worked on his part of the problem sharing rock sources and equipment.

While the solution to most diffuse pollution problems will be more complex than this, it is a good example of how concerted action can produce rapid results.

Text Box 6.4 Conservation tillage

Conservation tillage is a generic term used to describe systems of tillage or cultivation designed to conserve soil or water and to reduce their losses compared to conventional tillage systems.

The three main systems used in the USA are as follows (Carter, 1998)

Type of conservation	*Characteristics*	*Other terminology*
Mulch tillage	High percentage of crop residue on surface	Stubble mulch, trash farming, sod seeding, live mulch system.
Reduced tillage	Reduction in degree and/or intensity of tillage	Chisel plant, direct seeding, zero tillage, no-tillage, strip tillage
Ridge tillage	Use of raised beds in rows, often on contour.	

The factors which cause a farmer to adopt conservation tillage techniques are not always related to diffuse pollution or even soil erosion. For example, in many of the drier parts of the United States, conserving water is the main driver, while in the UK recently farmers have been adopting reduced tillage methods in order to save labour costs in establishing crops.

Nonetheless, whatever the reason, the effects can frequently, though not always, benefit the environment by reducing diffuse pollution. A summary of recent North American studies follows: (Carter, 1998)

Influence of conservation tillage (CT) on run-off and leaching parameters, compared to some form of conventional tillage, in recent North American studies (40 studies on eight soil texture classes from sandy loam to clay, (1991–96)

Effect of CT on measurement relative to conventional tillage % of studies

Measurement	*No. of studies*	*Decrease*	*No effect*	*Increase*
Runoff volume	16	38	38	25
Sediment in runoff	20	100	0	0
Sediment in runoff	13	69	8	23
Sediment in runoff	13	69	15	15
Sediment in runoff	7	71	29	0
Leaching volume	5	20	0	80
Nitrogen leached	13	36	27	36
Pesticide leached	11	8	8	85

Conservation tillage can be a useful tool in reducing the risk of diffuse pollution but it works best when combined with other techniques designed to optimize fertilizer and pesticide usage.

6.5 THE BRETAGNE EAU PURE PROGRAMME: BRITTANY, FRANCE

Like many other regions, Brittany has suffered from pollution of its rivers over many years: Initially this was caused by pollution from towns and industries, but agricultural pollution has also been a problem. One of the difficulties has been to manage a process that only produces effects in the medium and long term. Since 1994, the public institutions have worked together to co-ordinate their actions and to pool their technical, financial and personnel resources in programmes extending over several years (Le Gallic pers. com):

1995–1999: subsidized by more than 38 million euros
2000–2006: provisional subsidy is estimated at 91million euros.

These programmes, entitled 'Bretagne Eau Pure', are based upon:

- voluntary action and financial incentive
- an integrated approach to the problems of catchment areas
- application of regulations.

The 1994–1999 Bretagne Eau Pure programme has enabled the creation of a dynamic catchment area programme through the establishment of:

- advocates of catchment area projects: communities concerned with the quality of drinking water,
- professional agricultural committees on catchment area projects to advise farmers through meetings, field trials and investigations,
- a specialist group of agricultural technicians to advise about crop growing questions.

This has permitted:

- the start of guidance and practicable steps on the use of eco-friendly products and the reduction in quantity of chemical fertilizers,
- the encouragement of the users of fertilizers to meet standards of water quality more quickly,
- the adoption of eco-friendly products, resulting in the development of a pragmatic approach and training in the use of products to prevent run-off risks, bringing increased public awareness,
- farmers to adopt targets not based solely on commercial considerations,
- the achievement of the first measurable results that show an improvement in water quality with regard to the presence of particular plant treatments,
- the growth of demand for individual advice in preference to public awareness or association meetings.

The following points have emerged in the new programme for 2000–2006:

- the number of catchment area projects has more than doubled (from 20 to 43) covering more than a third of Brittany, 28% of farming and 58% of water resources.
- positive progress towards individual advice, with around 4000 cases to date,
- voluntary service advice in priority catchment areas in exchange for individual actions on the practicalities and delays in putting work into operation,
- the progressive move towards meeting advisory water targets on the basis of specifications guaranteeing the quality of services given.

All these actions and incentives from now on are accompanied and reinforced by regulatory actions. The project has found that regulation is essential, since 20 per cent of farmers did not respond to educational initiatives, even when supported by financial incentives. (Information from Henri-Claude Le Gallic, Manager of Bretagne Eau Pure, 4 Cours Raphaël Binet, 35000 Rennes, France).

6.6 THE USE OF WETLANDS TO REDUCE NITROGEN AND PHOSPHORUS POLLUTION FROM AGRICULTURE IN SWEDEN

6.6.1 The need for wetlands

Eutrophication effects along the Swedish west coast are associated with an increase in agriculturally-derived nutrient load, and include reduced oxygen concentrations in the bottom water of deep fiords and high production rates of opportunistic macroalgae, e.g. filamentous green algae like *Cladophora* and *Enteromorpha*.

One of the causes of this increase in the loss of nitrogen from Swedish arable land is believed to be associated with the drainage of lakes and wetlands. This has considerably decreased the retention capacity of nitrogen and is, along with the increase in gross load, believed to be one of the main causes of increased net load to the sea. The Swedish National Goal on the Discharge of Nitrogen is to reduce discharge to sea by 30 per cent by 2010, from a 1995 baseline. (Lann, pers. com).

Several initiatives are underway in Sweden to redress the problems. One of the earliest was the Halmstad project in the south-west, which by 1994 was responsible for the establishment of some 50 wetlands in the Halland region, including 20 in the municipality of Halmstad (Fleischer *et al.,* 1994, and popularized by the local authority publication *Wetlands for a Richer and Healthier Environment).*Wetlands are being recreated in urban and forestry landscapes there, as well as on farmland. Fleischer *et al*., (1994) describe the design details and individual characteristics, and reports on the early studies on nitrogen removal/retention in the new wetlands; denitrification is a principal removal process in the wetlands. Forest wetlands are seen as important for retention and loss by denitrification of atmospheric sources of NO_x. The project has been sponsored by the Swedish Environmental Protection Agency, and the municipality of Halmstad paid the construction costs. Other involved partners are the county agricultural society and administrative board, the limnological department of Lund University, and the division of water management at the Swedish Agricultural University in Uppsala. With passing years the work is becoming an important medium- to long-term study.

Elsewhere in Sweden similar approaches are being tried. In the County of Västra Götaland a range of different control measures is being sought, especially wetlands. Land-uses include arable farmland and forests, livestock include dairy cattle and pigs. There is a largely rural population of about 1.5 million. Monitoring of nitrogen and phosphorus in two streams in agricultural areas began in 1988, with a combination of grab-sampling once a week and continuously measuring of water-flow

Drainage from arable land is being collected into wetlands and ponds. The Swedish work has shown that it is not enough to construct new ponds; it is also necessary to restore old wetlands and transform ditches and channels to meandering streams (Lann, pers. com.).

6.6.2 Vänern Lake

This wetland project is focused on two watercourses that discharge into Vänern, the largest lake in Sweden. These watercourses have high levels of nitrogen and phosphorus, mainly from farming. In order to decrease these levels, the project aims to establish wetlands and buffer zones. The project started in August 1999 and continued for three years. The objectives were to establish 30 ha of wetlands and 140 ha of buffer zones and, by these actions, also increase biological diversity.

Participation from the farmers at an early stage was considered necessary for success. Farmer meetings were arranged where they were informed about the project. The farmers also filled in a form where they could indicate their interest in wetlands or buffer zones, and also if they wanted advice on plant nutrition in order to decrease leaching from farmland. Knowledge is a very important key. Obviously, not all farmers have a suitable area for wetland, but there are other actions that can be taken and in that way more farmers became involved. The advice was offered to the farmers free of charge.

It was attempted to find solutions that are individual at the farm level. This led to a close dialogue with the farmers who were able to contribute with their own ideas. The farmers thus participated in the process and influenced the actions. Farmers have received a subsidy when establishing wetland or buffer zones at the rates shown in Table 6.2.

Table 6.2 Subsidy rates for establishing wetland or buffer zones, Vänern Lake

Wetlands	230 to 300 euros per ha per year for up to 20 years
Buffer zones	300 euros per ha per year for 6 years
Total project cost:	890,000 euros.
Cost per wetland:	30,000–36,000 euros per wetland with a size of 15–30 ha
	10,000–12,000 euros per wetland with a size smaller than 1 ha

6.6.3 Results

The set objectives of the project were largely achieved. The approach has varied from farm to farm with some large wetlands being established, but also with smaller measures such as broadening or deepening ditches. The overall aim however was always to reduce the speed of the water so that nitrogen and phosphorus can remain in the wetland for a longer period of time before reaching the watercourse (Carlsson, pers. com). Other Swedish studies (Arheimer *et al.,* 2002) are suggesting, not surprisingly, that wetlands are unlikely to be a panacea for nutrient pollution, but that a suite of measures, able to be tailored to local conditions and circumstances are needed, as applied in the above examples where nutrient budgeting and waste management planning are also key elements.(see Chapter 2 and the following section).

6.7 APPLICATION OF BMPS AT THE FARM SCALE

The need for suites of measures to address rural diffuse pollution was indicated in Chapter 2 in the 'four-point focus' idea: measures need to be considered at the steading, water margins and in-field, as well as planning tools such as nutrient budgeting and farm wastes management plans (Figure 2.3). Although not always explicit in the case studies above, this approach applies at the farm as well as at the catchment scale.

Figure 6.2 schematically indicates the components of a typical example of a range of BMPs in practice. A number of measures have been employed to reduce the transfer of eroded soil and its attendant pollutants to a river. Action has been concentrated on critical areas with the agriculture over much of the catchment being unaffected. A train of BMPs has been employed ranging from conservation tillage through buffer strips to riparian wetlands. The figure recognizes that diffuse pollution often reduces to local 'hotspots' within the broad background level of contamination from land runoff. Measures need to be considered that can address both types of sources, since a lower level of input from a vastly larger area may be as significant as a local hotspot (e.g. a rill as compared to the loss from a multiplicity of field drains, for nutrient loss).

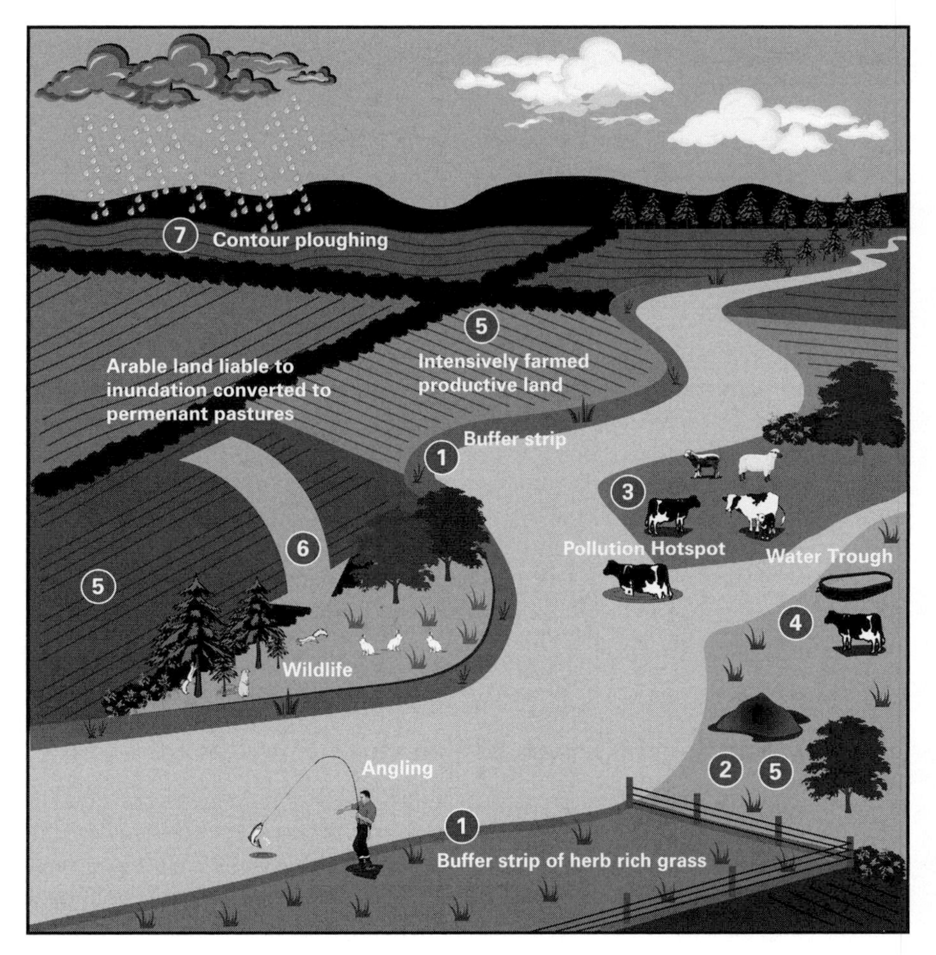

KEY

1. Riparian buffer strip: separating the farming activity from the watercourse
2. Farm waste management planning symbolized by field midden (dung heap): organic wastes such as slurry and manure need to be applied to farmland in accordance with a carefully drawn up plan to minimize pollution risk.
3. Example pollution hotspot; livestock collecting in one part of pasture and contaminating watercourse. Other hotspots are often farm gates at field/road boundary, and on arable land deep rills and gully formation in high risk erosion soils on sloping fields. See also (6) below.
4. Resolution of a livestock hotspot problem often involves fencing off direct access for livestock to a watercourse; an alternative supply must be provided for watering the livestock. Often a trough can be fed by gravity from the higher up the watercourse.
5. Nutrient budgeting is important on arable and livestock farms and seeks to match crop/animal needs with nutrients applied, taking into account season and soil type and other factors.
6. A part of a field liable to periodic major inundation and soil loss, will be less of a pollution and crop loss risk if converted to permanent pasture. Wildlife benefits should accrue.
7. Contour ploughing is one of several soil erosion risk management options for arable land in rolling landscape terrain.

Figure 6.2 schematic representation of a range of BMPs in practice.

Table 6.3 Lunan Lochs, Scotland: types of BMP measures needed and their application at farms in the catchment

Focus for farm measures	*No. of farms where measures sought (20 farms in the catchment)*
Steading	7
In-field measures	12
River margins	18
Management measures	20

Table 6.3 illustrates this for a study of 20 farms in the catchment of nutrient sensitive (mesotrophic) lochs in Scotland (Frost, unpublished report for Scottish Natural Heritage, 2003). At every farm management measures are appropriate, with other measures required according to circumstances.

6.8 CONCLUSIONS

All the above projects have features in common.

1. All involved the use of a range of best management practices focused on a defined environmental goal.
2. All involved the co-operation of a number of different bodies including government agencies and the farmers themselves.
3. All involved some element of financial assistance to help the farmers achieve the objectives being set.
4. All had a strong element of training for the farmers expected to implement the best management practices.
5. All involved the use of an environmental monitoring programme to check that the actions being taken were producing the desired results.

It is likely that all successful programmes will include the above five elements as part of their make-up.

LOCH LEVEN

Arable land in Loch leven catchment.

Swathe of influent river-borne sediment tracking shore of Loch Leven after heavy rain.

Soil erosion arable farmland, Loch Leven.

Free range pigs.

Riparian grass buffer strip.

Taking high risk (soil erosion) land out of arable production and establishing permanent pasture.

IOWA

Grass swale through arable land (Photo: Anne Sansom).

Dairy farmyard (Photo: Anne Sansom).

Feedlot (Photo Anne Sansom).

Nutrient Management and awareness raising. (Photo Anne Sansom).

7

Economic instruments, regulation and education

7.1 INTRODUCTION: PROXIMAL AND ULTIMATE CAUSES OF DIFFUSE POLLUTION

Diffuse pollution is pollution from the landscape. It has been increased by past developments to a point that in many places it represents a major threat to aquatic ecosystems. Examples of land-use activities that generate diffuse pollution are: installing paved surfaces (buildings, roads, parking lots) and storm and combined sewers to convey run-off from these surfaces; emitting pollutants into the atmosphere that deposit onto land; ploughing fields and applying chemicals on land; wetland drainage and in-filling; conversion of mangrove wetlands for aquaculture (shrimp production); deforestation and commercial afforestation; stream channel straightening and excavations causing stream bank erosion, stream corridor and floodplain development.

There are strong and complex linkages between land use policies and practices, macroeconomics of production and economic incentives, land degradation, and impairment of water resources. In solving the problems of diffuse pollution, distinguishing between the root causes of pollution and its symptoms is necessary (Table 7.1 and Figure 7.1). For example, anthropogenic eutrophication is a symptom of an increased input of nutrients from agricultural lands. But the root cause is land management practices that are not moderated by their impact on the aquatic environment. Environmental problems may reflect economic subsidies for fertilizers (UNEP, 1997). They can also be a consequence of the lack of failure to internalize the external cost of pollution and lack of enforceable pollution control laws.

Scientists and engineers have developed many control practices that help to minimize the symptoms of diffuse pollution – polluted run-off and degraded receiving waterbodies. But it is necessary to realize that diffuse pollution is a result of past and present social root causes, and technological failures and misdirections of the past. As an example consider a phenomenon of uncontrolled urban growth – urban sprawl. Rapid conversion of forests (deforestation) and agricultural lands into urban and suburban uses, continuous construction site erosion by building houses and roads in the countryside resulting in increased volumes of run-off that is far more polluted than in the pre-development stage, failing on-site sewage disposal (septic tank) systems, excessive fertilizer and pesticide losses from residential lawns,

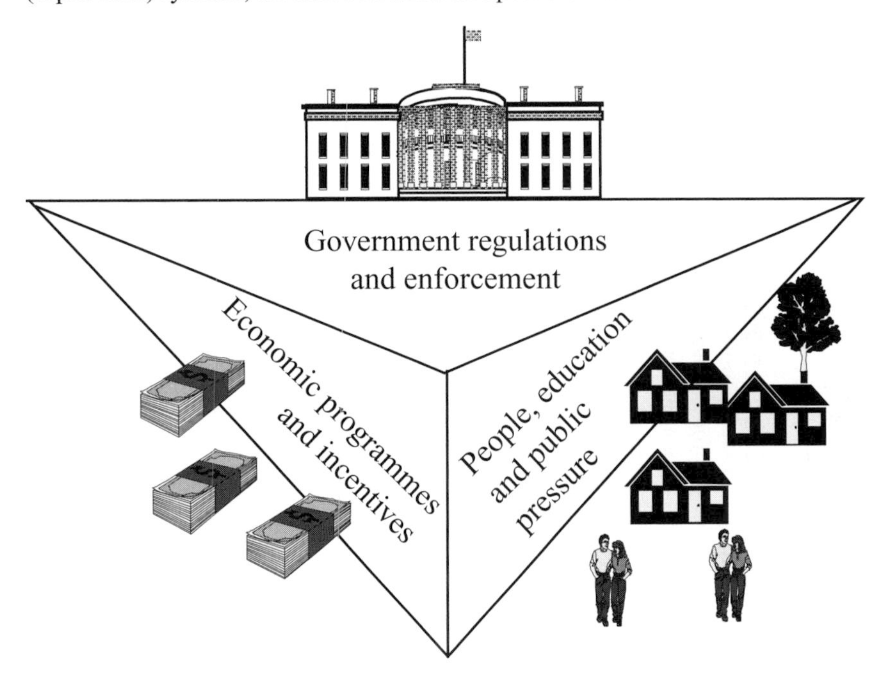

Figure 7.1 The relationship between the root causes of diffuse pollution and its symptoms means that solutions must be social, economic and technological.

Table 7.1 Examples of diffuse pollution problems and their root causes (modified from UNEP, 1997)

Pollution problem	*Root causes*		*Barriers to solution*
	Proximate	*Ultimate*	
Eutrophication and its symptoms (algae, low dissolved oxygen, turbidity, THM precursors).	Excess phosphorus and nitrogen inputs from point and non-point sources. Loss of the natural capacity of the system to assimilate and/or remove nutrients.	Population increase drives the demand for food. Subsidies for fertilizers and monocultural crops to farmers, international and domestic market for farm products, failure to value correctly the nutrient attenuation in wetlands when making decision to convert them to agriculture, lack of regulation banning phosphate detergents, lack of farmer incentives to implement best management practices, failure to internalize the damage cost into the cost of food and detergents.	Failure by authorities to evaluate economic consequences of eutrophication, emphasis on food production driven by population increases. Negligible influence of government environment ministries on agricultural and economic sectors. Public lack of awareness about the links between food production and deterioration of waters.
Siltation, turbidity and loss of habitat by increased sediment loads from diffuse sources.	Land use conversions, including deforestation and urbanization.	Population increase pressures and demand for new agricultural and urban lands, failure to include the cost of damages into land trans-actions, lack of land use planning and protection, demand for cheap wood and wood products.	Failure to develop long term policies on land use and sustainable development. Powerful vested interests in developed world controlling markets.

continued

Table 7.1 continued

Pollution problem	*Root causes*		*Barriers to solution*
	Proximate	*Ultimate*	
Loss of mangrove wetlands and coastal habitat degradation.	Rapid development of shrimp farms, direct use of mangrove wood for chips, competition for space in coastal areas due to development and land use conversion.	Massive coastal population growth resulting in pressure for land reclamation, pressure on developing countries to generate income by shrimp production, quick return due to the high price of shrimp, high demand for wood chips in east Asia, mangroves regarded as free commodities.	Failure to adopt integrated coastal zone management policies and failure to develop long term policies in land use and marine culture development. Absence of well-developed alternative economic activities.
Dissolved oxygen problem and poor quality of urban streams and coastal waters.	Discharges of un-treated or partially treated sewage, urban runoff with high BOD and toxic concentrations.	Population increases and emergence of shanty towns, poor or nonexistent sewerage and urban runoff collection systems, poor sanitation, traffic congestion in megacities and air pollution emissions.	Lack of funding for urban sewerage and sanitation infrastructure. Failure to implement and enforce environmental and health protection policies.

and increased pollution of receiving waters are the symptoms of diffuse pollution. The root causes of the pollution, however, are social, including the emergence of automobiles, deterioration of cities (notably in the USA) and people leaving the cities to sometimes very distant suburbs. In developing countries, the major root cause of diffuse pollution by rapid and uncontrolled urbanization is population growth (Novotny, 2000) and population migration in the cities.

Solving the pervasive diffuse pollution problem and improving the integrity of the receiving waters will require addressing both the symptoms as well as root causes of the problem. Diffuse pollution is an economic *externality.* Externality is an economic phenomenon in which an economic impact from upstream dischargers is transferred to downstream users of the resource by physical means (run-off, currents) and not by economic transactions and the downstream users damaged by this economic impact have no economic means to recover the cost of the damage. The solutions of the diffuse pollution problem must therefore be social, economical and technological (Novotny and Olem, 1994; Novotny, 2000) (see Figure 7.1).

Earlier sections of this book considered the available approaches to address proximal causes, and the following sections consider means to either directly address ultimate causes of diffuse pollution, or to provide mechanisms to deliver adequate uptake of best practice techniques to minimize pollution impacts.

7.2 ECONOMIC INSTRUMENTS

What constitutes an economic instrument in an environmental context (specifically the control of nitrogen and phosphorus pollution) was considered by Parr *et al.,* 1999, who argued that textbook definitions are generally flawed since they start with the notion that economic instruments are alternatives to 'command and control' regulations. In practice, economic instruments are often called upon to work within command and control structures to achieve specific outcomes by utilizing incentives to those involved. On this basis a much wider set of instruments can be classed as 'economic' than the traditionally discussed instruments such as taxation and tradable permits. It is useful to distinguish between those instruments in which 'the polluter pays', those in which the 'beneficiary pays', and those where there is cost sharing.

A further useful distinction is the point at which the instrument is applied. Analysing the point of application is aided by a consideration of the life cycle of the pollutant in question. A typical life cycle involves stages of production – sale – use/application – emission – pollution – damage. A tax, for example, could be applied at any stage in this cycle (Parr *et al.,* 1999)

The 'polluter pays' principle is often advocated as a basis for environmental economics. Wider socio-economic considerations need to be accounted for in the development and implementation of economic instruments, including ability to pay, and the social and economic costs of the consequences of continuing pollution. Economic instruments should lead to the development of more efficient, resource-conscious businesses. Vested interests may argue that if the instruments are too severe or punitive, they could potentially lead to the collapse of an entire industry, although experience with well-tried economic instruments such as punitive taxes on cigarettes or petrol to reduce use, suggests that is largely a theoretical construct. In reality, therefore, economic instruments attempt to strike the delicate balance of directing the overall cost of environmental protection towards polluters, whilst also trying to ensure the likely impact on the sector subject to such controls, is commensurate with the risks and is consistent with regulatory and other initiatives. A reasonable implementation timescale is also necessary. In other cases, for example in agri-environment support schemes, the 'beneficiary pays' principle is also relevant, where the beneficiary is, more often than not, the tax-paying public.

Table 7.2 summarizes the range of economic instruments which have been or are currently applied, with reference to nutrients (after Parr *et al.*, 1999).

Economic instruments therefore have two main purposes:

1. They shift the economic balance in favour of a desired behaviour change or other sought after outcome (e.g. improved environmental quality).
2. They can also be used for revenue-raising purposes. This tends not to be the case in wealthier countries (albeit consider the purposes to which money raised from vehicle excise duty and petrol sales is used), but in Eastern European/Eurasian countries, particularly, the use of economic instruments to fund government-run environmental authorities is receiving widespread interest. To date, this interest has focused on point sources of pollution (industry and municipal sewage treatment works), since these are easier to regulate. However, controlling point sources ultimately means that some diffuse pollution is also regulated, e.g. atmospheric deposition of NO_x and heavy metals, originally derived from industrial sources. Of course, the Eastern European/ Eurasian experience is not directly comparable to wealthier countries, and is born out of a severe economic crisis. The demise of the Soviet Union brought with it a collapse of both industry and agriculture – in some respects with environmental benefits. For example, by the end of the 1990s nitrate loads in the River Danube to the Black Sea were less than half of those a decade earlier (Parr *et al.*, 1999).

Economic instruments can include taxes that increase the cost to the polluter, for example a tax on the sale/use of nitrate fertilizer or pesticides. In theory, such taxes can deter profligate use and provide a stronger economic incentive to find alternatives. However, evidence that real reductions in use occur sufficient to meet

Table 7.2 Inventory of economic instruments (after Parr *et al.*, 1999)

Economic Instrument	*Comment*
Taxation	Any scheme where a charge is made on a product or activity. Terms like levy, charge and tax are generally interchangeable. Applied in a number of countries: Sweden, Austria, Netherlands, Belgium, Germany and Norway. Application is either at the manufacturer sales stage (Sweden) or on manure surpluses/on farm mineral balances (Netherlands, Belgium, France).
Liability (fines)	Fines (as opposed to taxes) on excessive nutrient application, as applied in Denmark.
Direct payments/ Subsidies	Payments made to agents to secure specific environmental goals. The incentive element is most important where schemes are voluntary. Applied in many countries (including the UK), Germany, Sweden and the US. Subsidies should include public provision of advice/assistance - which is widespread.
Cost sharing	Similar to direct payments but with contribution from the polluter. Applied widely in the US. Economic incentives are paramount in cost sharing bid procedures.
Tradeable rights	The allocation of rights to a specific pollution quota, e.g. catchment wide total maximum daily load. Main experience in the USA. (e.g. for Nox and Sox, also being developed for carbon trading)
Cross compliance	Typically where a non-environmentally-orientated direct payment is tied to environmental compliance. Applied notably in the US, EU and Switzerland.
Eco-labelling	Labelling of products that have been produced using environmentally appropriate methods and materials. Applied in Switzerland and the US (not nutrients). Also applied to some extent in other countries in terms of "organic farming".
Rights purchase	Similar to direct payments - involving the purchase of an agent's right to pollute and typically conservation banking of this right. Some experience in the US.

environmental needs, is equivocal. Anecdotal indications from Sweden (quoted in D'Arcy *et al.*, 1998) in respect of a tax on nitrate fertilizer, suggested that little benefit accrued. It was believed that perhaps the tax failed to achieve its environmental objectives because it was introduced shortly after a (bigger) increase in the cost to farmers of inorganic fertilizers, due to an increase in the price of oil (which affected the fertilizer manufacturers' costs). Swedish experience with economic instruments to reduce diffuse pollution associated with fertilizer application is reviewed by Kumm (1990) and Parr *et al.* (1999; see Text Box 7.1).

Table 7.3 illustrates the wider use of economic instruments. This table summarizes experience with economic instruments in several countries on the basis of a number of common themes:

- the objective sought in introducing the instrument;
- the nature of the instrument (tax, fine, subsidy, etc.);
- the coverage (nitrogen, phosphorus, both);
- the nature of the wider regulatory system within which the instrument operates;
- the level of application (e.g. amount of the charge or scale of application);
- issues related to administration, including costs etc.;
- the existence of complimentary measures; and
- an assessment of effectiveness.

An example of the clear failure of simple economic instruments that merely add to the price of a commodity (in the hope of reducing consumption sufficient to meet environmental targets) is the history of petrol (gasoline) price rises in the UK.

Text Box 7.1 Economic instruments and diffuse pollution associated with fertilizer application: a Swedish case study (Parr et al., *1999).*

Kumm (1990) discussed Sweden's efforts to control nutrient pollution from agriculture, which as well as fertilizer taxes includes intensive extension programmes and afforestation incentives. Although the programmes have raised revenues and been useful in local circumstances, it is argued that their efforts have been swamped by other policies which have tended to exacerbate the problem.

Some forms of charge on fertilizers have been used in Sweden since 1982. This began with a 'price regulation charge' and evolved into a tax. The original aim of the price regulation charge was to finance the reduction in surplus cereal production/export. The charge gradually increased and was 20 per cent of the price of fertilizers in 1990. In 1984 an environmental charge was introduced, which by 1990 was about 10 per cent of the price of fertilizers (in addition to the price regulation charge).

Text Box 7.1 continued

Kumm reports the difficulty of assessing the impact of the tax and charge given the other policies and trends occurring at the same time (such as that available for food protein levels which has encouraged fertilizer use and growing environmental awareness in the agricultural community).

The Swedish Environmental Protection Agency (1997) presented an evaluation of the taxation system used for commercial fertilizer. The purpose of the tax was to reduce demand and to fund a programme to reduce the impacts of agriculture on the environment. In 1994 the charge on phosphorus was abolished and replaced with a charge on cadmium. The original charge on cadmium applied to fertilizer with a content above 50g/tonne, but it was quickly adjusted when it became apparent that no commercial fertilizer had a cadmium content in excess of this limit. In 1995 the environmental charge was converted to tax status (the distinction being whether the revenues raised are hypothecated). While the 'price regulation charge' applied to nitrogen, phosphorus and potassium, the fertilizer tax applied only to nitrogen and phosphorus.

Originally administered by the National Board of Agriculture the tax is now administered by the Darlana County Tax Authority. The charge is levied on fertilizer manufactured in Sweden or imported. Manufacturers and importers are required to register, submit returns and pay the tax on the quantity they deliver each month. Of 45 registered manufacturers, 37 import for resale, 5 import for re-use and only 3 manufacture in Sweden. Some 'undeclared import for own use' is known to occur, but is not considered to be widespread.

The Swedish Environment Protection Agency (1997) estimated that revenues from theses charges amounted to SEK 300m (£23m – the vast majority of which comes from nitrogen). The percentage of price accounted for by the charge has varied over the period of implementation – at its height the charge represented 30 per cent to 35 per cent of the price, but later reduced to about 20 per cent. The tax is considered to be easy to administer. The National Board of Agriculture's administration costs were estimated to be SEK 0.5m annually (£400,000). The Agency (1997) suggested that the total costs amount to 0.8 per cent of revenue – assuming that the manufacturer's administration costs are equivalent to those of the authority.

It is estimated that the charges have probably reduced fertilizer use by 10 per cent to 20 per cent. In its evaluation of the tax, the National Board of Agriculture concluded that the main impact had been indirect – through the action programmes financed as a result of the tax. These programmes have continued despite the conversion of the charge into a tax (Parr *et al.*, 1999).

Table 7.3 Summary of applied economic instruments for the control of fertilizer application (Parr *et al.*, 1999).

Economic instrument	*Fertiliser taxation in Sweden*	*Direct payment with eco-labelling in Switzerland*	*Tradeable rights in US*	*Levy on mineral balances in Netherlands*
Objective	To reduce demand for chemical fertilisers and fund programmes	To secure public interest benefits from agriculture	To secure cost effective watershed improvements	Balance on farm inputs and outputs of minerals
Nature	Tax on manufacturer sales at point of delivery	Compliance with environmental objectives allows entry to premium organic markets	Varied. Some credit earned from point sources by securing improvements in alternative sources	Levy on surplus minerals, manure
Coverage	Predominantly nitrogen; some earlier coverage of phosphorus	Integrated	Nitrogen and phosphorus	Phosphorus directly but fertilisers (incl. nitrogen) generally
Wider regulatory system		Progressive direct payment system. Not part of common agricultural policy.	Direct payment, cost sharing	Agriculture sector targets.
Level	Represents 20% of the price of fertilizers. Has been higher.	40% of farms either environmental or organic farming	Trading activity generally limited	5 to 20 Dfl per kg of surplus
Administration	County tax Authority. Administration costs thought to be <1% of revenue raised	National	Local. National costs in setting up framework.	Regional authorities. Administration costs 'considerable'
Comple-mentary measures	programmes to promote extensive farming. Education and awareness programmes. Afforestation programmes.	Cross compliance. Direct payments	Cross compliance. Voluntary direct payments. Rights purchase.	Manure banks for disposal of surplus manure. Manure production rights (reducing quota with governments rights purchase and private trading). Manure supply standards. Farm relocation

Economic instrument	*Manure levy in Flanders (Belgium)*	*Nitrogen fines in Denmark*	*German water tax and compensation scheme*	*French pollution charges*
Objective	Sustainable mineral use on farms and meeting requirements of the EU Nitrates Directive	Better utilisation of Nitrogen in manure	Mainly drinking water protection (high reliance on groundwater)	Combat problems of intensive animal husbandry
Nature	Tax on surplus manure	Fines for excessive use of nitrogen. Effectively a quota.	Tax on abstraction used to compensate farmers for income loses	Tax on "net" pollution potential (reflecting abatement)
Coverage	Nitrogen and phosphorus	Nitrogen	Nitrogen and pesticides	Nitrogen and phosphorus
Wider regulatory system	Manure Decree. Manure Action Plan involving limits on the use of fertilizers.	Plan for Sustainable Agriculture		Subsidies for pollution abatement investment
level		Base exceedence charge combined with a per kg fine. 10 kg excessive use = £115	Lump sum payment based on standard assessment equivalent to £100 per hectare.	Payable on net pollution costs in excess of £700 per farm.
Administration	Region (Flanders)	National. High level of monitoring (spot checks) seems to be required (30% of farms)	State (federal). Requires substantial monitoring (soil nitrogen concentrations). Need for a large number of model farms.	Local river basin authorities. Administrative cumbersome because of need for farm environmental audits
Comple-mentary measures	Fertiliser use restrictions. Strict disposal rules and tiered quota for production at regional level. Development of manure processing facilities. Education and awareness.	Farm manure plans	Ecology program. Educational programmes. System of fines considered but not adopted.	Grants, soft loans, abatement rewards. Compulsory farm environmental audits.

Despite successive rises in price to the consumer of several hundred per cent over little more than a decade (due to increases in the price of crude oil as well as taxation) consumption has continued unabated. The punitive taxes may, of course, have slowed the rate of increase (by comparison with consumption in countries without such taxes, for example). European car manufacturers, however, have responded by creating ever more efficient motor vehicles that run further per litre of fuel. Motor vehicle traffic in 1999 was almost nine times that in 1950, and car traffic in particular increased by almost fifteenfold. In England during the last decade of the twentieth century, as petrol prices approached all-time maxima, traffic rose by 19 per cent on major roads (statistics from DETR, 2000).

By contrast, very substantial reductions in lead loadings into UK coastal waters were measured following the introduction of a tax differential favouring a new product – unleaded petrol – over the more heavily taxed traditional high octane fuel with lead anti-knock additives (alkyl lead). This was done prior to the use of leaded petrol being banned in the UK, USA and other countries, and prepared UK motor vehicle owners/users for that subsequent direct regulatory action.. *Both* fuel options were highly taxed by the UK government, but the few pence per litre *differential* was sufficient to reduce sales of the leaded petrol. Unleaded petrol was introduced in the UK in 1987, and since January 2000 only unleaded petrol has been sold. During that time lead emissions (from petrol-engine road vehicles) fell from 7.5 million tonnes in 1980 to 0.4 million in 1998.

One way in which taxation on a potential pollutant can be useful control mechanism is when the taxation income is hypothecated, i.e. ear-marked for use for environmental purposes. That has been the means whereby agri-support schemes have been funded in Sweden for example, where a small tax on agrochemicals financed advisory services to farmers.

Another type of economic instrument of interest is the concept of cross-compliance, which means that payments to producers for an output, are conditional on some other objectives also being achieved by the producer. Examples can be seen within the Common Agricultural Policy (CAP) of the European Union, where there has been a history of substantial production subsidies payable to farmers (Dwyer *et al.,* 2000, and Environment Agency 2002, Birdlife International 2002).

Table 7.4 shows how the nature of the crop a farmer chooses to grow, determines the pollution load from the farm. That decision is based on the economic climate within which the farm has to survive. Table 7.5 illustrates the different costs of pollution control measures that would be incurred by example UK farmers for largely off-farm benefits (D'Arcy and Frost, 2001). Cross-compliance mechanisms to require pollution control measures as a condition of receiving the crop production subsidy could be a means of overcoming the current economic disincentives to prevent pollution.

Table 7.4 Examples of variation in off field pollutant losses associated with different land-uses (D'Arcy and Frost, 2001).

Permanent grass	0.10 Kg TP/ha/yr	5% annual N application
Autumn-sown cereals	0.65 Kg TP/ha/yr	12% annual N application
Potatoes	0.80 Kg TP/ha/yr	20% annual N application
Brassicas	0.65 Kg TP/ha/yr	20% annual N application
Oil seed rape	0.65 Kg TP/ha/yr	30% annual N application

Notes
Nutrient loss coefficients, based on studies of arable fields in Windrush catchment (Johnes, 1996).
The land-use decision determines the probable incidental pollution load (diffuse pollution).

Table 7.5 Examples of estimated costs to farmers, for off-farm benefits, of controlling diffuse pollution. (D'Arcy and Frost, 2001).

Possible BMP measure	*Reduction in TP loss by soil erosion*	*Annual cost to farmer*
1. Convert to permanent grass	95%	£130/ha*
2. Contour ploughing	20%	£5/ha
3. Switch to spring sowing, not autumn	50%	£60/ha

Notes
P and sediment loss in this example, based on studies of target arable farms in Loch Leven Carchment, by Alan Frost and Ron Spiers of SAC.
*capital cost of becoming a livestock farm

An obvious alternative to cross-compliance, is to stop production subsidies and pay the money directly for the desirable environmental measures (direct payments, in Table 7.2). In theory at least, that presents no problem in many parts of the world (aside from cost to the taxpayer) but could in Europe, contravene the polluter pays principle. The latter concept argues that the polluter should bear the costs of preventing the pollution arising from his activities. That potential barrier to farm payments (for example) is surmountable by offering payments for broad environmental measures, such as landscape features with multi-purpose benefits, that typically include habitat creation, landscape enhancement, perhaps some public access etc., as well as playing a role in controlling diffuse pollution. In sensitive catchments, payments could be made for the land-use *decision* (e.g. with reference to Table 7.4, growing cereals or grass rather than potatoes would have a far greater impact in terms of reducing off-farm phosphorus loads than simply expecting the farmer to comply with pollution prevention best practice for potato growing. Pollution prevention best practices for each specific activity should remain outside

Levels of decision-making for pollution prevention in the rural environment

(a) Land-Use (Agri-environment)

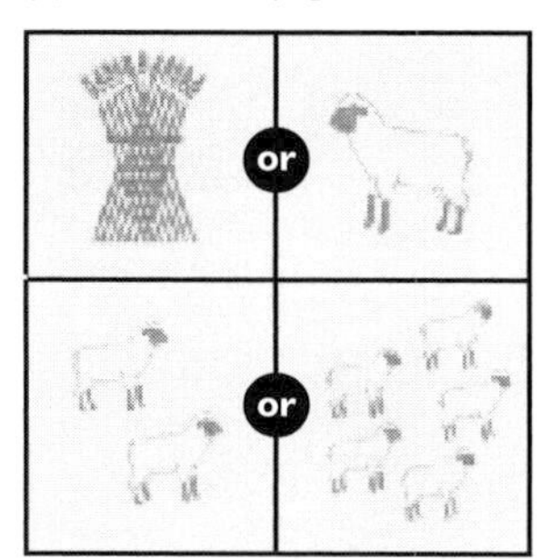

Example Decisions

- Crops or livestock or woodland?
- Autumn or spring ploughing?
- Livestock density?
- Conservation sillage?
- Landscape features?
  - e.g buffer zones (grass, hedges, woodland?)
  - e.g. wetlands

(b) Land Management (code of practice)

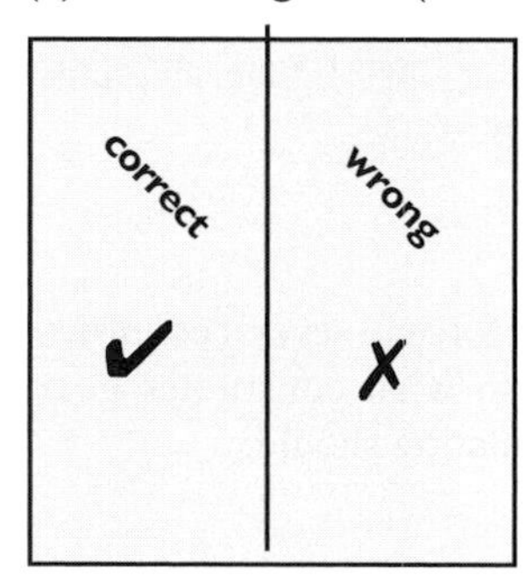

Example Management Actions

- Nutrient budgeting
- Waste management plan
- Non-application of specified pesticides with 6 m of a watercourse
- Disposal of pesticide rinse water onto crops

Figure 7.2 Levels of decision-making for pollution prevention in the rural environment; (a) could attract compensatory payments from the public purse, whilst (b) should be routine farming practice, without needing subsidy. Decisions on land-*use* (a) – for example arable or livestock? vegetables or cereals? free range pasture or intensive livestock? – often have more importance for likely loss of pollutants than trying to comply with notions of best practice for the chosen land-use (b). From D'Arcy and Frost, 2001.

subsidy support, in compliance with the polluter pays principle. These ideas are represented visually in Figure 7.2.

Frost *et al.,* (2002) examined six representative farms in Scotland to assess diffuse pollution and also habitats on the farms, with a view to considering how any modified agricultural support scheme could provide similar incomes to present schemes, assuming unchanging market circumstances. An example of the difficulties facing such efforts is evident in Figure 7.3, showing the range in income to the study farms and their varying dependence on subsidy.

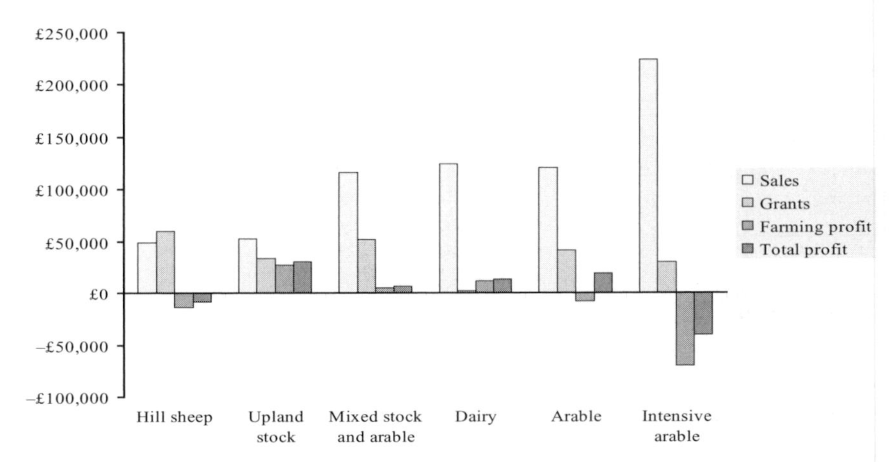

Figure 7.3 Sources of income and profit on the farms (Frost *et al.,* 2002). Note, all profits are before personal drawings. The figures are the mean of three years.

7.3 REGULATION

Before considering specific diffuse pollution measures in force in several countries, it is useful to consider the philosophical approaches to what constitutes good regulation (explored further in Text Box 7.2). Good regulation should:

- be effective
- be low cost
- involve minimal bureaucracy
- have regulated sector buy-in
- be simple and understandable.

The Scottish Environment Protection Agency considers these and additional ideas in some detail in a consultation process launched in September 2003 (SEPA, 2003), although saying little specifically with regard to diffuse pollution.

The nature of diffuse pollution (typically dispersed, individually minor sources, or true non-point sources such as fertilizer applications to soil), requires that legislation should apply at the land-use activity, rather than at any notional point of impact in a receiving waterbody. Therefore legal clauses that are enforceable for individually significant, major point sources, are generally inappropriate means of controlling diffuse pollution.

Regulations that are more appropriate for diffuse pollution either

- proscribe activities at the sources of potential pollution problems, or
- require prevention or mitigation measures associated with the land-use.

Text Box 7.2 What makes good regulation?

Desirable attributes and comments

Fair

Regulation should apply equally to equivalent polluting sectors, for example to agriculture and also forestry and urban development. (Equivalence means in terms of environmental impact.)

Simple

There has been a tendency in the UK, EU and USA to have a proliferation of environmental regulations as new statutes are developed apace with increasing awareness of environmental issues. In some instances even the regulators no longer know all the regulations, never mind the target sectors. Repealing antecedent legislation (incorporating equivalent provisions into new statutes as appropriate) is not always a straightforward option but may deliver the support of the regulated sectors.

Inexpensive

Regulation can be expensive to administer if too complex: for example multiple levels of approvals or permits, each requiring visits and separate paperwork, time-consuming delays for public consultation and appeal procedures, or individually drafted authorizations and conditions for target businesses. On the other hand, broader -based regulation with, for example, suites of standard conditions, risks over protecting the environment in some instances whilst not giving sufficient protection in others. But it does effectively reduce overall costs and bureaucracy.

Enforceable

Straightforward offence provisions are essential for effective regulation. But the regulatory agency must also adopt a positive (open, fair and predictable) attitude to carrying out enforcement to ensure compliance.

Targeted and proportional

Regulations should be in proportion to the risk and severity of environmental damage that might occur in the absence of regulation.

Examples of the former would be local authority byelaws making it an offence for householders to discharge spent engine oil or garden pesticide suspensions into road drains, or to license commercial businesses that wash householders' refuse bins such that they are not allowed to drain wash-waters (containing disinfectants) into highway drains or streams, but must recycle or otherwise safely dispose of such material. A UK example for agriculture is that some pesticides are only approved for use on farms on condition that they are not applied within a specified buffer zone width (e.g. 5m) from the top of the bank of a watercourse.

The second bullet point (regulatory provisions to prevent or control diffuse pollution) is exemplified by requirements to control pollution risks at source by enclosing oil and chemical tanks within containment areas (bunds) to provide interceptor systems for high risk areas, or to provide structural best management practices of the types described in earlier Chapters. Rural examples are buffer zones (see Chapters 2 and 5) – options to prevent or mitigate diffuse pollution in combination with other measures. They are widely advocated and are statutory measures in various circumstances in several countries (e.g. Denmark, see Text Box 7.3).

Text Box 7.3 Regulating the application of livestock slurries to agricultural land in the UK.

Under the terms of the Control of Pollution Act, 1974 in Scotland it is an offence to cause or knowingly permit pollution via the discharge of substances that are toxic or may be harmful to aquatic life. Whilst that is fine for regulating major point sources, such as municipal sewage effluent discharges and process effluents from manufacturing industry, it cannot be appropriate for diffuse pollution, since the best defence a polluter has (aside from an Agency permit) is to be able to show that the enforcing agency cannot prove his discharge caused, or was at least the major cause, of the measured pollution evident in the receiving waterbody. Successful prosecutions have been brought against individual farms when *gross* pollution has resulted from direct run-off into small watercourses (causing, for example, in-river BOD concentrations to be substantially elevated as compared to background values), but it is unusual.

The UK codes of good agricultural practice advise farmers that livestock slurry/manure should not be applied within 10 metres of a watercourse, and should not be applied during wet weather or onto lying snow. But historically in the UK such activities would not, in themselves, be offences. To effectively control diffuse pollution, *the bad practice should itself be an offence,* as is the case for the pollution risks inherent in the storage and handling of silage, slurry and agricultural fuel oil under UK regulations. For the future, a risk of on-the-spot fines – equivalent to parking or speeding ticket – might be more commensurate with the risks of pollution associated with bad practice, than formal notices and the preparation of cases to go before the law courts? The latter approach is more appropriate for major pollution events or for persistent refusal to adopt good practices.

Catchment-wide activity may not result in gross pollution incidents, yet may still result in elevated concentrations of pollutants, for example faecal pathogen concentrations in receiving watercourses. Kay *et al.,* (2000) and Aitken *et al.,* (2003) describe the evidence for intensive livestock farming adversely affecting the quality of recreational waters. Such impacts may not be obvious as gross pollution, as indicated for example by biological field scores (based on freshwater invertebrate fauna). Catchment-wide impacts (or effects on groundwater resources) are even more likely when nutrients are considered, especially if animal wastes are not taken into account by farmers when calculating nutrient budgets. Nutrient budgeting and farm waste planning are new skills for many farmers. Even where the need is accepted and there are significant economic incentives through consequent reductions in spending on inorganic fertilizers, farming practices still often lag behind environmental needs. Statutory measures are necessary, at least for the most sensitive and important at risk waterbodies. But they need to be supported by technical help, as used to be provided in the UK by government advisors at no cost to farmers, and as is still provided in many parts of USA and Canada (see case studies in following sections). The statute used for nitrate pollution in the UK and elsewhere in the European Union is the EU Nitrates Directive.

Example statutes to control diffuse pollution exist in various parts of the world. In the USA, the 1987 reauthorization of the Water Quality Act, added, in Section 101, to a declaration of goals and policy, the following principle:

> It is the national policy that programs for the control of non-point sources of pollution be developed and implemented in an expeditious manner so as to enable the goals of this Act to be met through the control of both point and non-point sources of pollution.

Section 319 of the 1987 Clean Water Act established a national programme to control non-point sources of pollution, but the focus of regulation for the target sectors is largely a matter for secondary legislation by individual states. Section 303 deals with water quality standards and implementation plans. The federal EPA is empowered to provide the states with funds for the control programmes and can withhold funding if an adequate programme is not developed by a state. In the USA, the Environmental Law Institute has published a useful *Almanac of Enforceable State Laws to Control Non-point Source Pollution* (Environmental Law Institute, 1998) that outlines out on a state-by-state basis the main statutes and their provisions, including offences. A useful discussion of regulatory aspects of urban diffuse pollution is also given in Horner *et al.,* 1994.

Urban drainage is controlled by the permit system in the provisions of Section 402. The EPA in America defines stormwater as storm water run-off, snow melt run-off, surface wash water associated with street cleaning or maintenance, infiltration, and drainage (Novotny, 2003). Permits for such discharges are issued by the states and are sought in respect of:

- stormwater discharges associated with industrial activity
- discharges from urban separate sewer systems, and
- discharges from construction sites.

Originally only urban settlements larger than 100,000 people had to comply with these requirements and construction sites greater than 2 hectares. But cities have used their prerogative to extend control to most suburban and nearby small urban areas, and many use this requirement to establish comprehensive drainage plans addressing water quality and flood control issues. The 1990 reauthorization of the Coastal Zone Management Act included specific provisions to address diffuse pollution, in recognition of the high proportion of the population living in the coastal zone and consequent threat to the environment. Mechanisms of implementation may include permit programmes, zoning, enforceable water quality standards, and environmental laws and prohibitions (Novotny, 2003).

Some states use economic measures alongside regulation, for example a penalty in proportion to the amount of impervious area drained by a business, has been developed in some towns and cities. An example is the Comprehensive Stormwater Management Plan (CSWMP) adopted by Eugene (Oregon) city council to bring together the various stormwater issues so that an efficient, co-ordinated and economical policy framework and management approach could be developed (Anderson, 1999). A similar integrated approach has been developed in the city of Atlanta, Georgia (Richards, 1999). Urban pollution and flooding issues are regulated in an integrated way in many parts of Europe too, notably at Malmö in south-west Sweden (Larsson and Karppa, 1999) and Dunfermline in the UK (McKissock *et al.,* 2001), but also increasingly in many other cities, amenity, landscape and habitat issues are also sought, usually by local planning regulations. Such integrated approaches are important and especially for drainage infrastructure. An integrated approach, including engagement with regulated sectors and the general public, will be important in Europe under the Water Framework Directive too (Chave, 2001 and Ellis *et al.,* 2002 with regard to urban drainage). But the control of pollutants at source, and of behaviours that can cause pollution is an equally important area for regulation (see Wilson, 2003).

For urban sources of diffuse pollution, a variety of statutory means of control exist (Table 7.2). Banning uses of problem persistent pollutants is considered in the Chapter 8 and the success of such action for lead in petrol has already been noted in Section 7.2 above. Such a regulatory approach has had success where the focus is the point of discharge for major point sources such as industrial process effluents. For diffuse inputs, controls need to apply at the point of sale (e.g leaded petrol) or manufacture (e.g. licensing for approved pesticides), or the point of application (usually restrictions on the types and categories of allowable use, e.g. for tri-butyl tin and other timber treatment chemicals and anti-fouling paints).

Regulating diffuse pollution in the rural environment is usually less comprehensive and often avoided due to political and economic concerns for the viability of parts of the agricultural sector. Examples are given below from European countries where a succession of EU Directives have required regulatory action by member states.

Table 7.6 Regulatory approaches to control of urban diffuse pollution.

Problem	*Example measures*	*Focus of regulation*
Herbicides in drinking water	Public health standards set by EU Directive (e.g European Commission 2001)	Point of consumption of water resource (water utility and government).
Herbicides in drinking water	Municipal policies set by towns and cities in Netherlands.	Policy to switch to non-chemical weed control
Lead pollution (air and water; associated with petrol engines)	Ban on use of lead in anti-knock petrol additives.	Sale of products.
Faecal pathogens	Local town council bye-laws e.g in Sweden, UK, USA.	Pet owner required to pick up and safely dispose of pet faeces
Polluting urban run-off (also flooding control)	National or state legislation requiring developers to provide BMP's (e.g Florida's stormwater rule of Florida Administrative Code)	Developers required to obtain stormwater permits.
Control of specific pollutants in municipal stormwater.	Federal USA : Clean Water Act. NPDES requirements for stormwater discharges.	Municipalities (173 US cities and 47 countries, initially)

Note

NPDES = National Pollutant Discharge Elimination System, extended to cover stormwater by the 1987 amendment of the Clean Water Act, which prohibits the discharge of any pollutant to navigable waters from a point source unless authorized by a NPDES permit (Cook, in Herricks, 1995).

Text Box 7.4 Case study on control of diffuse pollution: Denmark (from Farmer et al.*, 2000).*

a) Codes of good agricultural practice (CoGAP)

The requirements for good agricultural practice in Denmark are set out in environmental regulation. Denmark has a comprehensive environmental regulatory framework and a number of specific regulations which impinge upon agricultural practices and require standards above and beyond EU Directive requirements. These regulations are aimed at minimizing diffuse pollution and directing farmers towards increasingly sustainable farm practices. The most important environmental regulations, in terms of the control of diffuse pollution, are detailed below, as outlined in the Danish CoGAP. The Code is structured so that there are five sets of precautionary measures.

1 Precautionary measures regarding the treatment of manure. This includes the law on the use of fertilizer and cover of vegetation in agricultural holdings, (No. 472, July 1 1998, also implementing the Nitrate Directive). This regulation defines a maximum nitrogen quota at the farm level, which the overall use of nitrogen must not exceed, entails rules on crop rotation, utilization of livestock manure and catch and green crops. Thus, farms have specific quotas of nitrogen indicating the maximum amount of nitrogen to be used on the farm. To this end, even the utilization of nitrogen in manure is established. Farmers are also obliged to make crop rotation plans, field and fertilization plans or 'accounts' (for which punctual submission to a government body is essential on an annual basis). The regulation also seeks to ensure the careful management of manure and fertilizer through a system which registers the transfer of manure and fertilizers to farm yards. Finally, the regulation also requires farmers to ensure they hold 6 per cent catch crops and 65 per cent vegetation cover all year. Ministerial Orders add further detail to this framework, regulating the location of new domestic livestock holdings in the open land, the location of open dung hills, and the storage of liquid manure all to minimize the risk of pollution. These restrictions are valid for farms with an income of greater than DKK 20000 and farms with more than 10 animal units, with a density of animals exceeding 1 animal unit per hectare and if the farm receives greater than 25 tons of manure. Precautionary measures regarding the treatment of manure. Ministerial Order No 877, on livestock manure and silage etc, regulates the location, construction, storage and capacity requirements for manure and silage. It also regulates the ratio between the production of manure and area available for the spread of manure on farmland. Thereby setting standards for the density of livestock, which in general are

1.7 LU per hectare for pigs and 2.1 LU per hectare for cattle. The maximum densities will be reduced to 1.4 LU for pigs and 1.7 for cattle by 1 August 2002.Precautionary measures when applying and spraying pesticides, including technical requirements and safety inspections of spraying equipment. In terms of pesticide application, under Law No 266 of May 1993, farmers are required to allow inspection of spraying equipment, to maintain detailed journals of pesticide use, identifying fields, their size, crop, types of pesticide used, quantity and date of application.

2 Precautionary measures to protect rivers and lakes, in terms of pollution and sedimentation. Under the River Conservation Act, Section 69, farmers are required to leave uncultivated borders of 2m strips along both sides of all water courses and lakes larger than 100 m^2, in all rural areas. These borders or riparian zones should be kept clear of any crops, plantations, modifications of the terrain or enclosures. The purpose of this is to protect the water in rivers and lakes against pollution from soil erosion and leaching.

3 Precautionary measures regarding sludge and sewage content. Under Ministerial Orders No 49 & 56 / 2000, concerning the use of sewage sludge, all sludge must be tested to ensure that it does not exceed certain heavy metal concentrations, agents injurious to the environment. There are also storage capacity requirements, etc.

Controls and on-the-spot checks under the rural development programme will be used to observe any infringement of the codes of good agricultural practice. Penalties and sanctions will be imposed according to the severity of the infringement, which can include complete discontinuance of support and refund of support already paid out (Farmer *et al.*).

Other examples of regulatory measures for controlling diffuse pollution in agriculture are given in Farmer *et al.,* 2000.

The importance of regulation in the cost-effective achievement of environmental objectives has recently been highlighted in studies in the Netherlands (Zeegers *et al.,* 2002) and in Canada (Tuininga, 2002). Tuininga concludes that a combination of voluntary and mandatory enforcement measures have proven to be an efficient means to bolster environmental stewardship in the cattle industry in Ontario, supported by a commitment to open communication by all parties. Quoting Boyd (1998) advocating tough regulation as a motivator, Tuininga argues that mandatory compliance achieved through enforcement may effectively reduce the level of farmer to non-farmer conflict in Ontario. In Alberta, Canadian regulators have a sequential approach to carrying out enforcement options, including a number of measures that would be useful statutory powers for diffuse pollution. The seven enforcement stages reported (in D'Arcy, 1998a) by the Pollution Control Division of Alberta Environment Protection (AEP) are:

1. Warning – a record of an environmental contravention, typically minor.
2. Ticket – a fine levied for a minor offence.
3. Enforcement Order (EO) – an order from the Director of Pollution Control requiring specific remedial action or information
4. Administrative penalty – a penalty levied by the Director of Pollution Control for an environmental offence.
5. Prosecution – an enforcement action carried out through the criminal court system.
6. Court Order – an order from the civil court requiring specific remedial action.
7. Cancellation of approvals or certificates – a withdrawal of the offending party's permission or certificate of qualification to operate.

The idea of a ticket as a regulatory response to a failure to follow best practice, is commensurate with the diffuse pollution concept – the individually minor yet collectively significant; an idea worth exploring.

Finally, it is important that regulation across sectors and between states is even-handed and does not present unfair competitive disadvantages. This is often not achieved (for example the existence of oil storage regulations for farmers, but not for small industrial businesses in the UK in the 1990s) or, at best, evidence is unclear (Williams *et al* 2002). MacDonald and Williams (2001) compared regulation in Denmark, Portugal and Scotland, albeit not for diffuse pollution but for the tanning industry. It is unlikely their findings would be very different for urban drainage, or application of agrochemicals, management of manure and slurry, or storage and handling of chemicals and wastes – since the focus of regulation is so variable and the approach to enforcement is so much a matter of national culture and tradition. The more relevant findings of Macdonald and Williams for this discussion concluded that within the EC there were differences in:

- the degree of decentralization of the regulatory process
- inspectorate training and resources
- inspection frequencies and inspection criteria
- the use of enforcement mechanisms and the levels of fines imposed, and
- the costs to industry associated with regulation.

These discrepancies make it even more important that regulation is supported by education initiatives with regulated sector buy-in, and that the economic climate within which the regulated sectors operate is in alignment with the aims of government as indicated in regulation.

7.4 EDUCATION

The USA has adopted an extensive regulatory approach, at federal and state level, to address diffuse pollution from urban and industrial sources (see Table 7.2 above). For agriculture, by contrast, a voluntary rather than a regulatory approach to preventing diffuse pollution has been developed. Although resource intensive, the voluntary approach is encouraging cooperation and goodwill in a proactive and educational way. There have been studies which indicated that the majority of farmers were happy with this approach and were well aware that a failure to meet targets voluntarily might mean more regulation in the future (Sansom, 1999a and see Text Box 7.5). It is not uncommon to see regulation and education as stand-alone alternatives in possible control strategies (e.g. Robinson, 2000); this is further considered in section 7.5.

The educational approach requires:

- clear explanation of the importance of the problem
- explanation of the technical measures needed to control the problem
- unanimity of message by all organizations disseminating advice to the target sector.

The achievement of these requirements is expensive (see the Brittany project referred to below and earlier in Chapter 6, plus references to funding programmes in USA projects). An aim is to achieve target sector buy-in (conceptually, if not literally) to their role in the problem

The USA approach favours targeted catchments, and similar approaches are prevalent elsewhere, reflecting the nature of diffuse pollution problems: most clearly manifest at a catchment level and involving many contributing sources, whose collective impact needs to be addressed. In France a comparable high resource educational approach is being developed in Brittany (see Chapter 6) where 250 million francs was spent on educational programmes and support measures between 1995 and 1999, with a massive 600 million francs approved for 2000-2006 (Le Gallic, pers. comm.). But all those actions and incentives are accompanied and reinforced by regulatory actions. The project has found that regulation is essential, since 20 per cent of farmers did not respond to educational initiatives, even when supported by financial incentives. It will be interesting to see how effective are new educational initiatives such as that in the UK for pesticides, led by the pesticide manufacturers and retailers (see Text Box 7.6). Lessons from elsewhere suggest that education alone is unlikely to meet environmental needs.

Text Box 7.5 Educational approaches to controlling diffuse pollution from agriculture in the USA (Sansom, 1999a)

The relationship between regulators and individual farmers in the USA is not close. The Environmental Protection Agency (EPA) tends to utilize the Natural Resource Conservation Service (NRCS) and the Soil Conservation Districts to work directly with farmers on soil and water problems. These organizations have little or no regulatory responsibility and are trusted by the farming community. The state and federal EPAs are only involved in serious pollution incidents. Farmers had varying reactions to regulation, but most understood that protecting the environment and good farming practice made good business sense. Some sectors of the agricultural industry have worked directly with regulators to improve their business and standing with the consumer. In some states the farming community has been asked to design the regulation its own industry needs to comply with environmental targets.

Farmers have a close working relationship with the NRCS and the Soil Conservation Districts that goes back to the 1930s. The Soil Conservation Districts are run by a democratically elected local group of people including farmers. The Districts employ 'District Conservationists', farm advisers with backgrounds or training in agriculture and/or the environment. More recently other organizations have been employing farm advisers and the most notable of these is the Farm Bureau (a farmers' union) in Utah. The aim of the Farm Bureau is to speed up the voluntary approach and try to avoid the need for regulation. The Farm Bureau recognized that for this to work the farmers must trust the adviser and that it was best, therefore, if the adviser was with the Farm Bureau itself. The different government and non-government agencies were all very aware that farmers learn first from other farmers. Agencies such as the NRCS and the Soil Conservation Districts would facilitate setting up demonstration farms and organizing farm walks and talks that were run by the farmers for other farmers, with the agencies taking a secondary role. In Iowa the farmers themselves had taken this a step further and set up their own organization, the Practical Farmers of Iowa (PFI), that now has 500 members. The PFI's mission is to promote farming systems that are profitable, ecologically sound and good for families and communities. The organization carries out its own research and shares information with members. Once again the key word is trust. The PFI has been successful because farmers tend to trust each other more than they trust anyone else.

The USA government and the EPA have been promoting 'Community-based Environmental Protection' 'aggressively"' for many years, and the Clean Water Action Plan in 1997 gave the promotion additional impetus and resources. Projects are usually based on catchments and involve all the local stakeholders by helping them to identify environmental problems, set priorities, find solutions and make the necessary changes. The projects are results-orientated and led by local people with the government and non-government agencies facilitating them. The whole process focuses on the benefits for stakeholders and landowners, the local economy and the environment. This approach encourages a sense of local ownership and achievement that leads to long-term support and responsibility. The EPA were frequent leaders in this field, supporting projects with staff and resources and producing a range of materials to help local communities.

The USA places a high priority on education and communication, particularly at the early stages of a project. Figures of 10 per cent of a project's budget spent on communication and education were not unusual and many people said that the results were worth it. The Cooperative Extension Service provided all the materials needed to get messages across and worked closely with the facilitators and local people. There was also a strong emphasis on education for younger generations, with the result that individuals are now taking more responsibility for their local area.

Text Box 7.6 The Voluntary Initiative: a best practice initiative for use of pesticides.

The Voluntary Initiative is a UK partnership venture to reduce pollution risks associated with use of pesticides, and was developed by the pesticide industry in response to proposals for establishment of a pesticides tax in the UK. The project partnership comprises representatives of stakeholder groups such as the Crop Protection Association, National Farmers' Union (NFU), NFU Scotland, Ulster Farmers' Union, Country Landowners Association (CLA), AEA, NMC, UKASTA and Water UK. Leaflets advocating good practice have been produced for pesticide users, and highlight eight essentials to protect water as follows:

1. Planning: know where the water is on your farm and where it drains to
2. Consider whether using a pesticide is necessary in the first place
3. Talk to your agronomist about the products you use, their risk to water and how to keep them out of water
4. Clean up any spills or splashes immediately
5. Pick your filling site with care-about 40 per cent of contamination by pesticides is caused by run-off from farmyards
6. Maintain your sprayer properly; stop leaks and drips
7. When spraying, keep well away from watercourses, use buffer strips and prevent drift
8. Clean everything carefully afterwards and dispose of wastes safely and legally.

Every drop counts is the slogan of the campaign. Working with the regulators, the effectiveness or otherwise of the initiative is being evaluated in nineteen target catchments across the UK.

It is obviously helpful to any educational initiative to be able to demonstrate benefits to participants; if economic advantages can accrue so much the better. The latter have been demonstrated in farm waste minimization initiatives in the UK, where pesticide targeting and nutrient budgeting have achieved tangible cost savings for farmers (Nicholson and Baldwin. 1999, Baldwin *et al.,* 2003). Surprisingly large cost savings can be achieved by farmers – surprising in that theoretical economic argument would suggest they should be achieved without the need for educational initiatives by environmental state agencies (see Text Box 7.7 documenting the Big Spring Project in Iowa). In practice there are often cultural or other traditions that are barriers to simple market forces.

Educational programmes can also be cost effective at a state or province level. Text Box 7.8 refers to the comparison made by New York City between the cost of paying farmers to minimize diffuse pollution in the two catchments that comprise the potable supply for the city, and building and operating high level treatment facilities. Similar approaches have been taken in Australia, for urban run-off, to protect and ensure the continuing supply of water resources.

Text Box 7.7 Iowa case study (Sansom, 1999a)

Big Spring in the northeast, which produces 15,000 gallons of groundwater every minute from 100 square miles of agricultural catchment. During the 1960s and 1970s the use of nitrogen increased almost threefold, increasing the concentrations at Big Spring by a similar amount. By the early 1980s the concentration of nitrogen commonly approached the limit set by the US Environmental Protection Agency for drinking water (45 mg/l). The amount of nitrate flowing annually from the basin in surface and groundwater was equivalent to a third of the nitrogen fertilizer applied by farmers. Estimates of other losses of nitrate, such as uptake by aquatic plants, suggested that the total loss was equivalent to half the nitrogen applied.

These findings led to the creation in 1987 of the Big Spring Basin Demonstration Project, a cooperative effort involving farmers and state, federal and local agencies. The project made use of funds available under s. 319 of the US Clean Water Act, 1987, to implement education and demonstration programmes to improve the economic and environmental performance of agricultural practices, and to expand the scope of water quality monitoring. Between 1981 and 1993, as farmers became more confident that lower inputs of fertilizer would still work, nitrogen input declined from 79 to 52 kg per acre, a reduction of 34 per cent, with no effect on yields. The total annual saving to farmers in the region was about $360,000.

Text Box 7.8 New York City case study (Sansom, 1999a)

New York State law authorizes New York City to regulate the two upstate catchments – the Croton and the Catskill-Delaware – that provide the city's drinking water. These two catchments contain parts of eight counties, one city, 60 towns and 11 villages, and more than 500 agricultural and horticultural units. The two catchments together produce 1.2 billion gallons of water daily, providing drinking water for nine million people. In the past the water quality has been exceptionally high, but in recent years land development and increasing agricultural intensification has caused the water in the Croton reservoirs to become seriously degraded. This deterioration has triggered an EPA requirement that filters must be installed to prevent microbial contamination. The Croton system produces only about 10 per cent of the water consumed by New York City. If similar degradation were to occur in the more remote Catskill-Delaware system, and filters became necessary, the costs would be enormous. Construction costs could exceed $5 billion and annual operating costs would be approximately $300 million.

In 1993, the EPA agreed to postpone the decision on filtering the Catskill-Delaware supply while the city attempted to demonstrate that it could maintain the quality of the water though catchment control.

Following this a comprehensive 'Whole Farm Planning Program' was established and endorsed by the New York State agencies, New York City, the farmers and the EPA. The programme assumed that in a well-populated rural area, well-managed agriculture was the best protection for water quality. Regulation was unlikely to succeed with independent farmers and the programme had to be a voluntary one based on providing incentives to farmers to participate.

New York City provided $4 million for the first two years of the programme during which the objective was to develop plans on 10 demonstration farms. Of the roughly 500 farms in the catchments, the majority are dairy farms and the 10 selected as demonstration farms were all dairy. The second phase of the programme is now under way and involves implementing Whole Farm Planning on 85 per cent of the remaining farms, with funding set at $35 million. Community planning is also being demonstrated with six towns assuming a pilot role to look at wastewater, storm water and drainage. These programmes should enable New York City to continue to protect its water supply at a fraction of the cost of a filtration plant.

Further details in USEPA (1997) *People, Place and Partnerships: A Progress Report on Community-based Environmental Protection* (USEPA, Washington, DC)

Any education campaign should include a means of assessing its effectiveness. A professional evaluation requires pre-awareness and behaviour surveys prior to the new initiative or other educational activity (Usman, 2000) and this is especially important for diffuse pollution where there are many individuals collectively comprising the polluters. At least some initial idea of success criteria is important for educational efforts too; although often lacking, even in well-planned and well-executed campaigns. Usman *et al.,* (2000) compared proactive educational initiatives in the UK and Canada, in relation to oil pollution – a major issue for urban catchments especially.

7.5 INTEGRATING APPROACHES TO THE PREVENTION AND CONTROL OF DIFFUSE POLLUTION

The same factors represented in Figure 7.1 as the context within which diffuse pollution occurs, are also the key to resolving diffuse pollution. Thus the statutory framework establishes the requirements for the various land-use sectors to adopt prevention and control measures. The economic environment is structured to be at least compatible with those requirements and preferably supportive. And irrespective of the balance between those two, an education effort is essential to explain the statutory requirements and also the details of pollution prevention best practices. There is not, therefore, a real choice between regulation or advice; regulation is merely an element in the array of persuasion techniques to favour pollution prevention and control practices.

Figure 7.4 shows trends in water quality in two small agricultural streams in the Forth catchment in central Scotland, where a similar approach to solving chronic poor quality was adopted, but with different outcomes.

A partnership was effected with local agricultural advisory organizations, and joint meetings held to which all farmers in the problem catchments were invited. The same format was followed in both catchments. The pollution problem was explained by the regulator (pesticides) and the technical solutions were offered by the agricultural partners, including follow-up visits over succeeding months. In the Dreel Burn water quality improved, but in the West Peffer comparable progress was not evident. It may be significant that a willingness to be receptive to the friendly partnership approach (i.e. to change practices) was greater in the Dreel catchment because the regulator was fortunate enough to catch good evidence for a prosecution just prior to the launch of the initiative there. During the period of the study that was not the case in the other stream.

The Swedish experience of economic instruments to influence nutrient use (see Text Box 7.1) highlights the importance of the overall economic climate in shaping

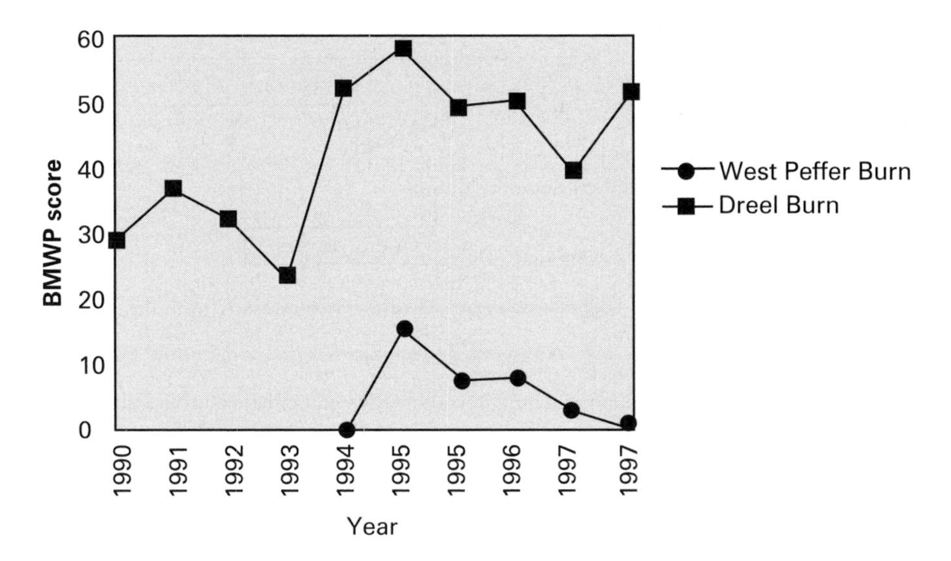

Figure 7.4 Water quality trends in two small Scottish agricultural streams (from D'Arcy *et al.,* 1998).

behaviours, and anecdotal indications suggest that regulations that go against that economic climate will be ignored or incur confrontation between regulator and regulated sector. A simplistic indication of the relationships between these three key factors (education, regulation and economic climate) is given in Figure 7.5 Success cannot be secured where any one of the factors in Figure 7.5 is missing. For example in parts of the USA huge amounts of money have been spent on voluntary schemes and many individual successes achieved – but, in the absence of a regulatory regime, a few uncooperative farmers in a catchment can ruin those achievements by not being interested in participating in such voluntary projects.

An example of the lack of synergy between regulation and economic environment is the effort in the UK to control pollution risks associated with the storage and handling of pollutants. In 1974 the Control of Pollution Act published provisions for regulations to control the risks, but no general regulations were produced. Instead, eventually the Pollution Prevention (Silage, Slurry and Agricultural Fuel Oil) Regulations 1991, were brought into force, controlling only a tiny percentage of spillage risks. On any example farmyard, adjacent oil tanks (for example heating

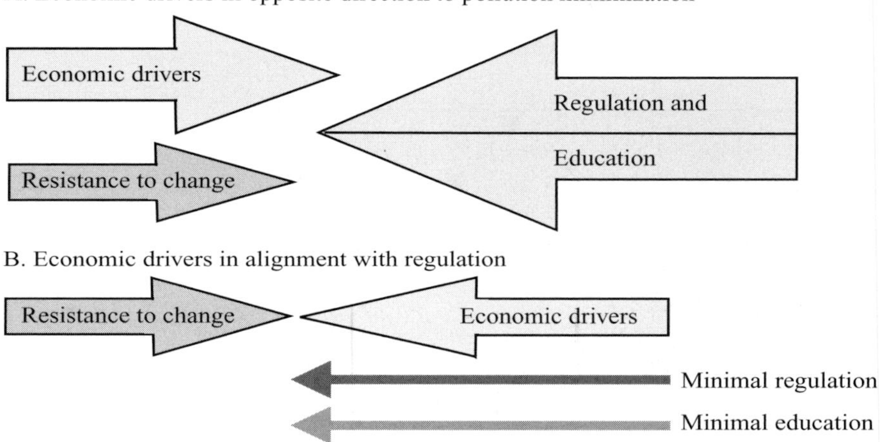

Figure 7.5 How to achieve minimal regulation: alternative combinations of measures to control diffuse pollution.

oil alongside agricultural fuel oil) were subject to different controls, the former being unbunded and the latter subject to regulations requiring bunding. Fuel oil stores for non-agricultural businesses in the same area were not regulated, thereby putting the farming industry at a disadvantage compared with other businesses. It also undermined the efforts of storage tank manufacturers to establish an adequately sized market for low risk (integrally bunded) tanks. It is therefore still possible, and substantially cheaper, to buy unbunded storage tanks. More general regulations (but only for oil) eventually came into effect in October 2001. That example illustrates a failure to recognize the importance of economic factors in harness with regulation (in this instance the fragmentation of a market for a non-polluting product). It also illustrates how educational efforts can also be made more expensive and difficult as regulatory positions fragment, and inconsistencies are all too evident.

There is often a belief that polluting sectors are hostile to regulation and therefore non-regulatory options are the only way forward. This view is prevalent in many parts of North America in respect of the agricultural sector (e.g. Iowa case study see Text Box 7.9).

The Iowa example of a well-funded partnership approach to pollution prevention is rightly acclaimed (see Text Box 7.9) for its engagement with the target sector and funding of the initiative by local taxation of agro-chemicals. Nevertheless, it remains to be seen if it can achieve more substantial reductions in contamination than reported to date.

Discussions with practising farmers, often suggest a much broader range of views, many recognizing the need to protect environmental resources, and the facts of

Text Box 7.9 The voluntary approach in Iowa (Sansom, 1999a)

Ninety per cent of Iowa is farmland, with 75 per cent planted with crops such as maize and soybeans. Farmers in Iowa spend about one billion dollars every year on farm chemicals, especially nitrogen fertilizers and herbicides such as atrazine. The use of these chemicals has doubled since the early 1960s and created significant environmental problems. Crop yields doubled between 1960 and 1980. About 130,000 Iowans drink water from private wells that have nitrate levels that exceed federal health advisory limits. During the summer of 1991 the public water supplies in Des Moines and Iowa City were periodically declared unfit for infants. In 1987 the Iowa Groundwater Act was passed providing a non-regulatory approach and funding in the form of subsidies, education, technical assistance and research to reduce the use of agro-chemicals. Ten years later, although the diffuse pollution problem had not been solved, the use of inorganic fertilizers had decreased by 15 per cent.

The non-regulatory approach has been an administrative success. The collaborative approach and the emphasis on non-regulatory tools allowed state officials to implement an extensive programme that involved all relevant public agencies and several non-profit organizations. Initially, the chemical manufacturers and the chemical dealers refused to admit there was a problem, but their confrontational tactics backfired and they were brought on board later on.

The approach has been a political success too. The legislators emphasized that the diffuse pollution problem was a civic issue, for which all industries and all citizens share some of the responsibility. Despite keen opposition from well-funded industrial lobbyists, the legislators were able to build a broad coalition of public interest that won through. This coalition attracted federal dollars at a time when most other budgets were being cut. The state and its leaders have won wide praise and several awards for their efforts.

The future is unclear, but fees on chemicals in Iowa still yield $3 million a year to help pay for the education programme. If it is possible to farm profitably by reducing the use of chemicals, perhaps these programmes and market forces will solve a good part of the diffuse pollution problem. If targets are not met, the possibility of regulation remains, but by then the majority of farmers should already be in compliance.

Further details in J. De Witt (1994) 'Civic environmentalism – alternatives to regulation in states and communities' Congressional Quarterly Inc.

human nature that require a regulatory under-pinning of established desirable behaviours. For example, a Nuffield Farming Scholarship study in 1999 in the USA reported as follows:

> In general, farmers do not trust federal or state interference and would rather be independent and rely on the market for income. Having said that, farmers do not refuse federal or state funding in the form of cost-share programmes (75 per cent funding) when it is offered! The reactions to legislation of the farmers spoken to varied considerably. Some were resigned to legislation, seeing it as a necessary evil. Others thought it was fair provided that all sectors of society were regulated and took their share of the responsibility for a catchment. Some farmers felt that they were being unfairly targeted. Others believed regulation was necessary to stop the 'cowboys' that were giving the industry a bad name. Flexibility was also important, with the voluntary approach being far more flexible than the regulatory one. A dairy farmer from Vermont said that a nutrient budgeting exercise had saved his business. A farmer in North Carolina said that he liked regulation! The improvements that had been made to the tracks on his dairy farm had made his operation cleaner and easier to manage and he was very grateful. The majority of the farmers visited understood that environmental protection and good farming practice made good business sense. One farmer from North Carolina said he would like more information or evidence that the work he and other farmers were doing was benefiting water quality and the catchment. Others felt that the industry needed more recognition for its efforts
> (Sansom, 1999a).

In conclusion it is clear that an integrated approach to controlling diffuse pollution is essential and achievable. In the longer term it is most likely to be cost-effective if it encompasses the three key elements for control:

Education: encompassing *the need* for prevention/control actions; *how to* prevent and control pollution; and knowledge that *it is an offence* to cause pollution.

Regulation: establishing basic statutory measures to require compliance even by those minority parts of polluting sectors who would not otherwise participate and, equally important, establishing a consensus position on what is the minimal acceptable level of environmental protection.

Compatible economic environment: there should not be economic factors working against the prevention of pollution. New economic measures may be appropriate to align business practices with regulation and education.

These are the three parallel strands of environmental protection and pollution control.

8

Diffuse pollution and sustainable development

8.1 INTRODUCTION

The most widely quoted definition of sustainable development is based on the following passage from *Our Common Future,* by the World Commission on Environment and Development (Brundtland Commission, 1987):

> Humanity has the ability to make development sustainable ... to ensure that it meets the needs of the present without compromising the ability of future generations to meet their own needs.

Adger, in O'Riordan (1995) notes that the central aspects of that definition are the focus on the environmental basis for human activity, and the time dimensions of development and well-being, The long-term, cumulative impacts of human activities is a fundamental concern.

This chapter first looks at where diffuse pollution may be compromising the ability of future generations to meet their own needs. Trends in the impacts that were characterized and exemplified in Chapter 1 are considered, and a few classic examples of unsustainable land uses are outlined. Progress, where documented, is also noted. Whilst best management practice techniques, as discussed in earlier chapters, are essential specific measures to address pollution problems on the ground in a variety of situations, they cannot be the whole story. Additional measures are necessary – for example, more general approaches to specific pollution problems by limiting use of certain problem substances, or the development of clean industrial processes and other technologies that do not generate problem pollutants. More than that, as discussed in some detail in Chapter 7, there is a broad range of more fundamental means of addressing diffuse pollution – looking at ultimate causality, and the scope for regulation, economic incentives and education to foster more sustainable practices. These last considerations are the domain of the wider perspective on environmental issues in a socio-economic context that is at the heart of sustainability. Several examples are given in subsequent sections of integrated approaches to environmental issues (e.g. in relation to agriculture in Section 8.4, and urban development in Section 8.5).

Figure 8.1 encapsulates how an integrated approach to solving diffuse pollution problems can been adopted (based on example initiatives with agricultural sector in Scotland, and also on experience with implementation of new development practices for urban drainage systems (D'Arcy and Harley, 2002). Clear identification and characterization of the environmental problem is the duty of the environmental bodies. But their remit then extends to working closely with the various sectors involved, to better understand the problem (including identification of economic drivers involved and institutional and behavioural barriers to be addressed). The most appropriate, workable and acceptable solutions to the problem can then be co-developed with the sector and co-promoted with them.

In the UK, a much-quoted, but rarely exemplified, political phrase from the turn of the millennium, is very apposite to diffuse pollution control and prevention: *the need for 'joined-up thinking'*. Some of the means discussed in the previous chapter of persuading polluting sectors to adopt best practice to prevent pollution, may be complemented by pollution control agencies and other concerned organizations making common cause with other environmental bodies, that are also concerned about different, but related, environmental consequences of development. Key environmental issues that offer common cause opportunities are:

- loss of habitat for birds and other wildlife in intensively farmed land
- increased risk of flooding if undeveloped rural land is drained for agriculture or forestry, or if conventional drainage technology is used for new urban development
- loss of wetlands and other natural and semi-natural habitats
- the need for informal recreation and public amenity features in new developments (urban, and also rural such as commercial forestry)

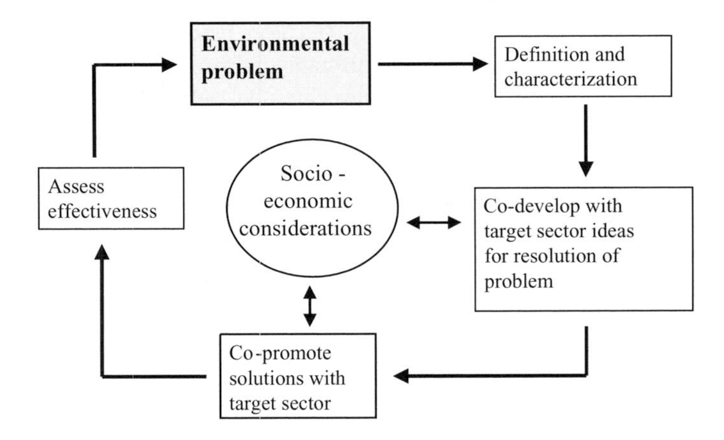

Figure 8.1 An integrated approach to resolving environmental problems

- reduced dry weather flows in watercourses after urbanization, with adverse consequences for stream fauna and flora
- urban traffic and public health/global warming

Similarly there are economic imperatives that offer opportunities for environmental improvements that include diffuse pollution control, for example:

- restoration of contaminated land for development
- support for farmers and rural populations in general
- improvements to public transport services with consequent curtailing of the environmental impact of cars.

By seeking links between problems and solutions, cost-effective actions can be identified (with multiple benefits), that together can lead to development that is more sustainable than current practices.

Furthermore, as discussed in the previous chapter, the root causes of diffuse pollution – social and economic drivers – need to be addressed, by a combination of education, economic instruments and regulation.

8.2 TRENDS

It has long been recognized that diffuse sources are the most important causes of pollution in the USA (USEPA, 2000), and the agricultural sector along with urban run-off are the biggest problems (Weitman, 1996). There are significant economic impacts. For example, in some New York City water supply reservoirs watersheds are degraded by extensive development (Croton Reservoirs) and agriculture (Delaware Reservoir) resulting in significant diffuse pollution problems and eutrophication (Division of Drinking Water Quality Control, 1993). Even with the lead taken in the USA, developing the best management practices philosophy to control diffuse pollution during the 1980s, reversing trends has proved very difficult.

SEPA, the Scottish Environment Protection Agency, has predicted that diffuse pollution from agriculture will become the most significant water pollution problem in Scotland by 2010 (SEPA, 1999). A similar position is believed to be likely in England and Wales. Although the rate of increase of pollutant concentrations in some rivers has tailed off in recent years, (see Figure 8.2 which shows the rising trend for nitrate in a Scottish river) reversing ecological impacts may not be so straightforward, especially for lentic waters. It is even more problematic for corresponding groundwater resources, where concentrations reflect build-up over decades and can be expected to take decades to decline. Agrawal *et al.,* (1999) note that chemical and bacterial treatment of groundwater for nitrate removal relies on advanced technology and is considered costly even in the developed world; in a country like India where economic resources are inadequate, action to address diffuse agricultural pollution that is endangering potable groundwater resources for future generations is urgent.

For the UK as a whole (D'Arcy *et al.*, 2000a), major impacts of agricultural practices are anticipated, associated with nutrient enrichment: phosphorus in inland waters and nitrate in coastal waters. Particularly for naturally nutrient poor waters, unsustainable agricultural practices (and, historically, forestry too) jeopardize the nutrient status of those precious resources. Nitrate is also a major issue as regards concerns for drinking water resources. Over the next 20 years, in order to comply with EU drinking water standards, the estimated cost for raw potable supplies for the UK water industry, has been estimated to total £199 million (Skinner *et al.,* 1997).

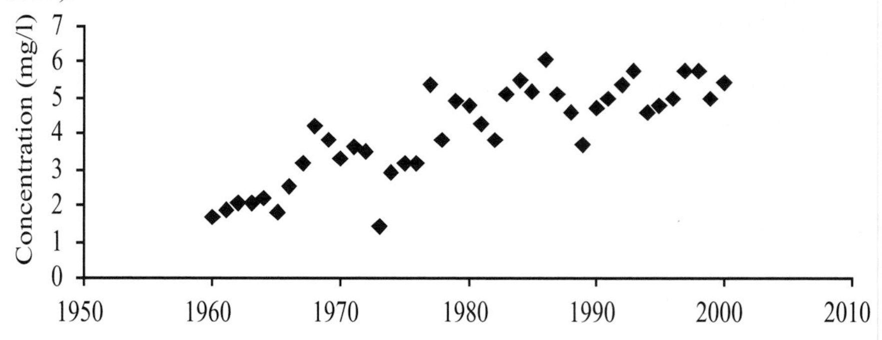

Figure 8.2 Concentration of nitrate as NO_3-N in River Tyne, Scotland (SEPA unpublished data, 2001)

One of the saddest examples of the unsustainable nature of many land-use activities is the decline in water quality evident in the Venice Lagoon (Bendoricchio *et al.,* 1999, 2002). Indeed, the example of Venice illustrates a number of sustainable development issues, not just diffuse pollution (see Text Box) 8.1.

The earlier problems with major point source industrial and municipal sewage effluents, have been exacerbated by over enrichment associated with agriculture in the catchment (Sfriso *et al.,* 1988, Bendoricchio *et al.,* 1999). The problem was exacerbated by the Common Agricultural Policy of the European Union, which historically supported intensive agriculture, with insufficient requirements for mitigating measures to protect water resources. But progress is being made: between 1991 and 2000, the annual nitrogen load from the drainage basin decreased from 9,000 tonnes to 6,500 tonnes per year. To minimize risks of nuisance algal blooms however, a goal has been set to reduce nitrogen loads below 3,000 tonnes per year.

In Germany diffuse pollution is also recognized as a major concern, associated with unsustainable practices. It has been estimated that 60 per cent of German soils are sufficiently or excessively supplied with phosphorus, (Mohaupt *et al.,* 2000). In the last two decades a sharp decline has been reached in surface water pollution from the nutrient components ammonia and total phosphorus, due to installation or upgrading of waste-water treatment plants and also, in the case of phosphorus, the increasing use of phosphate-free detergents. Nitrate levels however, remained high. In 2000 the target quality class of II or better was reached for ammonia at 42 per cent of 151 stations, for total phosphorus at only 24 per cent of 151 stations, and for total nitrogen at only 13per cent of 151 stations. Diffuse sources from agriculture are increasingly coming to dominate in nitrogen pollution. They have fallen by only 15 per cent in the last decade, and phosphorus loads from agricultural land have even risen slightly (7 per cent). Spatial distribution reveals the largest diffuse sources in areas with high livestock densities and soils which are prone to leaching or run-off, as is the case with phosphorus in the peat soils of the far north-west of Germany. Similarly, the whole of the north-west (sandy soil) and some areas near the Alps produce the highest nitrogen loads.

As regards trace metals, German studies have shown that unsustainable practices continue to prevent the achievement of water quality targets on many rivers, despite reductions of up to 95 per cent in loads from major point sources. It was recommended that the use of heavy metals in construction and road traffic should be tested for their necessity and alternatives be found (Mohaupt *et al.,* 2000). Otherwise, contamination by those persistent pollutants will continue.

Another diffuse source of toxic metals pollution is abandoned mine waters; the problem is likely to get worse before it gets better. The full effects of decades of oxygen having access to hitherto anaerobic rock strata containing significant quantities of iron pyrites, oxidizing it to soluble ferrous sulphate, will become

Text Box 8.1 The Venice Lagoon – a case study in sustainable development issues

The ancient city of Venice is a cultural and historical treasure of international renown, yet the city itself is threatened by unsustainable environmental consequences of development, including global warming. Many sources of greenhouse gases are diffuse in origin: methane from livestock, carbon dioxide from combustion of fossil fuels in motor transport and many domestic heating systems, as well as garden fires and solid fuel fires for cooking and heating homes. Rising sea levels already threaten priceless art treasures and the very existence of Venice. The fate of the city is not just important for its inhabitants and well-wishers; it is a massive foreign exchange earner for the region as tourists are attracted from all over the world. The wider environmental issues that impact on Venice include abstraction of groundwater and natural gas resources in the region (implicated as causes of instability in the lagoon) as well as discharges of sewage and industrial effluents into the Venice Lagoon. Eutrophication of the Venice Lagoon has become a major issue, with agricultural diffuse sources of nutrients being especially important (field applications of up to 250 kg N per hectare; Bendoricchio *et al.,* 2002).

The following information is from the Master Plan 2000, of the Veneto Region in Italy, to take action for pollution abatement in the Venice Lagoon and its drainage basin.

Venice Lagoon: total surface 540 km^2, canals and deep water areas 66 km^2, shallow water subtidal areas 243 km^2, intertidal areas 98 km^2, salt marshes 11 km^2, enclosed fish farming areas 92 km^2, islands 29 km^2, mean water depth 154 cm, total water volume 600 million m^3, mean tidal excursion at the three inlets 60cm, mean water volume exchanged with the Adriatic sea 450 million m^3/day, mean freshwater volume from the drainage basin 2.6 million m^3/day.

Drainage basin: total surface 1,878 km^2, inhabitants 950,000, people working in industry 160,000, agricultural surface 1,300 km^2, livestock (head of cattle and pigs) 250,000. Comprises catchments of Sile and Adige rivers, and cities of Venice, Padova and Treviso.

Nutrient load reduction from diffuse sources: Economic measures – subsidies will be paid to farmers to switch to crops and cultivation methods that result in a lower environmental impact.

increasingly evident as water tables rebound, because of cessation of pumping at abandoned mines. Depending on local hydrogeological conditions, ferruginous springs may appear where formerly unpolluted waters used to seep from the ground or, where groundwater recharges the local watercourses, riverbeds may become orange over a broad area of land, often some distance from the former mineworkings. It can be years before water tables rebound and polluting springs appear, and then it is decades before there is much prospect of substantial improvement as indicated in Figure 8.3 (contributed for this book by Paul Younger of Newcastle University; see also Younger, 2000). Comparable problems exist in many parts of the world (e.g in South Africa, see Heath *et al.*, 2002).

Odell (2001) notes that the most threatening environments (for human life) are those of cities in Africa, Latin America and Asia. Whilst there has been, for example, a five-fold increase in urbanization that has occurred in Africa since 1950, the problem is not confined to the developing world. The populations of Buenos Aires and Rio de Janeiro are stabilizing or declining, rather than increasing further. But Dallas and Las Vegas in the USA, have grown faster than Calcutta and Mexico City. Urbanization is a global issue.

The notoriety of Los Angeles for smog well illustrates the problem of traffic in urban centres, contributing to contaminated road run-off, as well as more widely recognized air pollution. Los Angeles is one of the world's fastest growing cities (Odell, 2001). The growth in road traffic is a source of pollution wherever there are cities. It is a sustainable development indicator for the UK and unfortunately is still a rising trend (see Figure 8.4). The diffuse pollution concerns are not just from particulates, NO_x, SO_x and partially combusted hydrocarbons washed out of the air

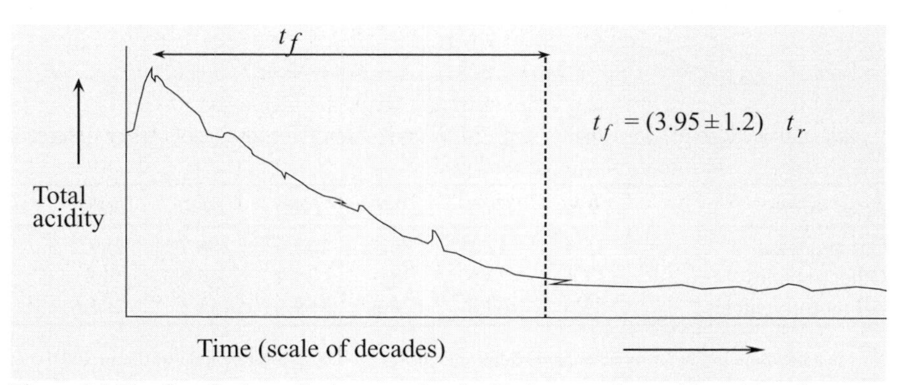

Figure 8.3 The first flush: pollution associated with former mineworkings. Decades are likely to elapse before peak concentrations of pollutants decline in outbreaks of contaminated water associated with abandoned mine workings. Source: Paul Younger, Department of Civil and Environmental engineering, University of Newcastle, UK.

or washed off highways and other urban surfaces after dry deposition. Persistent pollutants, such as toxic metals arising from the use of these materials in brake linings and tyres, for example, need to be reduced at source by changes in the motor vehicle components (specification of persistent pollutants).

In general, there can be little doubt that diffuse pollution impacts are amongst the most serious consequences of unsustainable activity i.e. economic activity that threatens the well-being of future generations. Although improvements are now becoming evident on many fronts, the trends suggest that many of the remaining problems will deteriorate before they improve.

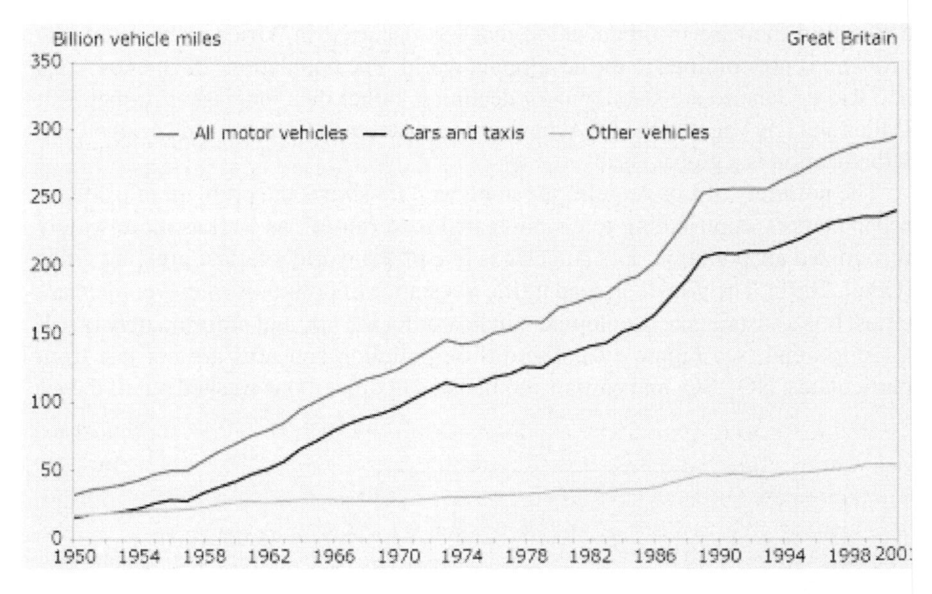

Road traffic	*1950*	*1960*	*1970*	*1980*	*1990*	*2001*
Car and taxis	15.9	42.3	96.3	133.6	208.7	239.7
Other vehicles	17.1	27.5	28.3	35.4	46.6	54.9
All motor vehicles	33.0	69.8	124.6	169.0	255.3	294.6

This is a headline indicator of sustainable development for the UK. Motor vehicle traffic in 2001 was almost nine times that in 1950, and car traffic, in particular, has increased by fifteen times. Overall, traffic levels rose by 1.4 per cent between 2000 and 2001; car and taxis traffic was up by 1.9 per cent whilst goods vehicle traffic remained almost unchanged on 2000.

Figure 8.4 Road traffic, by type of vehicle, 1950–2001 (source: DfT)

8.3 RESTRICTING THE USE OF IDENTIFIED PROBLEM CHEMICALS

Some pollutants – typically chemicals, often manufactured for a particular industry – are found to have such severe adverse environmental impacts associated with their use that the most sustainable (and practical) means to control the pollution problems are to ban or severely restrict the use of those products or constituents. The first class of chemicals to be recognized in this way were the organo-chlorine pesticides developed in the post-war decades to assist with the need to feed the growing global population: DDT, and the 'drins' – aldrin, dieldrin and endrin. The first popular assessment of the adverse impacts of such pesticides was Rachael Carson's classic, *Silent Spring* (Carson, 1962). Subsequently, retrospective studies in various parts of the world have largely confirmed the suggested impact of organo-chlorine pesticides on wildlife in areas where they were applied, for example in the UK, the declines in populations of birds of prey (see reviews in Brown, 1976, and Sheail, 1985) and others (Mason and MacDonald, 1993). The bio-accumulation of persistent pesticides or their breakdown products in animal tissues resulted in far-reaching impacts, beyond the target species, including possible health risks to humans at the top of many food chains, if widespread use continued. The widespread use of a few toxic substances as pesticides is also unsustainable because of the probability that target species develop resistance to the chemicals (e.g. Gupta, 1983). Reviews by Henriques *et al.* (1997) and Gupta (1983) provide interesting developing world perspectives on agrochemical pollution, including more recent generations of pest control products. Heriques *et al.* review human health impacts as well as ecological ones.

Returning to organo-chlorine substances, concerns were raised, at almost the same time as for pesticides, about a constituent of some industrial compounds; polychlorinated biphenyls, or PCBs. Widely used in transformer oil, spills, drips and leaks from a vast number of local sources on business parks, industrial sites and other premises all over the world, as well as transport by air following combustion of contaminated materials, have resulted in widespread pollution. As a result, PCBs have been found in the fatty tissues of fish and especially of predatory birds and mammals from many countries. PCBs have been implicated in ill health or even local population declines for a variety of predator species that are exposed to bioaccumulations of these persistent organo-chlorine additives via food chains, (Sheail, 1985, Kannan *et al.,* 2000, Helander *et al*., 2002, and see Text Box 8.2). Environmental concentrations of these substances have also declined gradually since they too were phased out (Alcock *et al*., 2002). Lake sediments may be contaminant sinks that could still present pollution risks. In several respects, PCBs represent a classic instance of diffuse pollution – contamination arising from the widespread

Text Box 8.2: Polychlorinated biphenyls (PCBs) – a diffuse pollution history

- PCBs were first used in 1929, in the USA, originally manufactured by Monsanto Chemicals and later by others, and introduced into Europe about ten years later.
- Uses were mainly as protective coatings, plasticizers, sealers in water-proofing, printing inks, synthetic adhesives, and in hydraulic fluids, thermostats, cutting oils, grinding fluids and electrical equipment.
- Attractive properties for industrial use – chemical stability (non-degradable chlorinated hydrocarbons) and insolubility in water (but lipophilic) – also made them potentially dangerous as environmental pollutants.
- Initially detected as substantial background contaminants; organo-chlorine compounds present in the tissues of animals being analysed for pesticides such as DDT and dieldrin, (e.g. present in pike, *Esox lucius,* from Skane in southern Sweden, but low concentrations in pike from unpolluted Lapland). The same 'mystery' substances were then found when analysing feathers from white-tailed eagles, *Haliaeetus albicilla*. Samples were taken from specimens collected since 1888, but the chemicals were only present from 1942 onwards, suggesting they could not be metabolites of chlorinated hydrocarbon pesticides, which only came into use in Sweden after 1945. A corpse of a white-tailed eagle from near Stockholm contained enormous amounts of the unknown compounds, which were finally identified as polychlorinated biphenyls. Meanwhile, PCBs were also reported in the UK, and elsewhere.
- In 1970 the extent of contamination by PCBs of British wildlife was reported: 196 livers analysed from 33 species, and 363 eggs from 28 species of bird, obtained between 1966 and 1968. PCBs found in terrestrial species from most regions, and in most eggs from freshwater species. Residues also present in seabird eggs from one west coast and two east coast colonies. Highest liver values found in freshwater fish-eating birds and bird-feeding raptors. Two herons (*Ardea cinerea*) had the highest recorded residues for PCBs anywhere in the world. Experimental work confirmed the toxicity of PCBs: significant, but less than dieldrin or DDT pesticides.

  In the Irish Sea PCBs were implicated in a major seabird population crash in autumn 1969 (Holdgate 1970).
- Speculation as to sources of contamination considered the following:
  - via chimney stacks of industrial plants
  - as constituents of industrial effluents
  - through the gradual wear and tear of products containing PCBs

- as a result of incineration of discarded products containing PCBs
- contaminated sewage sludge disposed at sea.

Contamination was not a consequence of a major accident at sea, or flagrantly polluting industrial effluent discharges, but was a function of the uses and disposal of PCB materials. *Diffuse pollution* emerged as a useful concept and a real and important issue.

- From 1970 Monsanto stopped supplying PCBs for 'non-controllable' uses such as the manufacture of lacquers, paints, lubricants and paper. Only remaining permissible uses is in closed systems: dielectric fluids in transformers and capacitors, but this was prohibited in systems manufactured after the middle of 1986. In 1996, EU Directive 96/59/EC on the disposal of PCBs (and polychlorinated terphenyls, PCTs) was adopted, requiring equipment containing PCBs to be registered and safely disposed of; the environmental focus is now on safe disposal of ageing sources of contamination.

(After Sheail, 1985, with updates as referenced.)

use of a class of substances that gain entry to the aquatic environment via scouring of spills, leaks and other individually minor sources, into surface drainage networks or groundwaters, and via atmospheric deposition after combustion of materials.

Taking the Canadian Arctic as an example, useful reviews of impacts and levels of contamination of persistent pollutants, and example pathways for contamination are given in Braune *et al.,* 1999 and Macdonald *et al.,* (2000).

In a perhaps more sustainable approach to the difficulties of pesticides contaminating water resources, in the Netherlands a community engagement process is being undertaken to secure public acceptance for the presence of some weeds within paved areas of towns and cities. One of the community campaigns involves the slogan: 'just because it is green doesn't mean it has to be killed'.

In the Dutch province of Zeeland the majority of the local and provincial authorities have signed an agreement in which they declare to stop using herbicides in 2005 on roads, in parks and on their own grounds. By 2003 they will have reached a 75 per cent reduction compared to 1999 (Hollemans, 2002). The environmental and economic need is to maintain the quality of raw water supplies, jeopardized by diffuse (mainly urban) sources of pesticides (Bannink, 2002). To that end, a handbook for municipalities has been developed, to move away from use of herbicides, to an alternative strategic view of weed issues in the municipalities (Baardwijk and Spijker, 2002).

That process of education and engagement with the public has the potential to lead to more sustainable solutions than searching for technical solutions such as new classes of herbicides (to replace banned products), or more efficient technology for stripping herbicides from raw water supplies. Bannink (2002) also argues that conventional approaches present water consumers with the bill for highly technical, expensive purification techniques, but the bill should be met by the people benefiting from the use of herbicides on pavements. The most sustainable solution, he argues, is to end the emissions to surface water by developing alternative weed control/ tolerance practices.

In contrast to pesticides, for which there are influential national regulatory bodies that require evidence of research to provide some confidence that new chemicals should not have adverse impacts on non-target species, there is generally no such mechanism for industrial chemicals (e.g. PCBs). Furthermore, a new class of environmental concerns has become apparent in recent years: the properties of a range of manufactured substances that allow them to behave as hormone analogues in the aquatic environment (see IEH (1999), CES (1993) and Tyler (1998) for reviews of the problem). Initial research has focused on the presence of these substances in municipal sewage effluents, but it is likely that diffuse sources are also important (Ellis, 1996; Marttinen, 2002). Imposex, associated with antifouling paints with active ingredients such as tri-butyl tin (Bauer *et al.,* 1997; Bright and Ellis, 1989), is also evident in estuarine and inshore habitats; as for PCBs and organo-chlorine pesticides, the regulatory option has been to restrict or ban the uses of these seriously polluting substances.

Alkyl forms of toxic metals (such as the anti-fouling paints noted above, which are also widely used in timber treatment) also are often directly toxic. Other forms of toxic metals can also be a concern too, of course. Mohaupt *et al.* (2000) state that the situation in urban stormwater management calls for urgent action to minimize diffuse sources for heavy metals. As well as control measures such as settlement ponds and infiltration systems, the need for the continuing use of heavy metals in the construction and the motor industry needs to be tested, and if possible, alternatives found. That action would be the ideal, truly sustainable option for preventing pollution by these substances.

In the Netherlands toxic metal pollution associated with the use of zinc and copper in the construction industry (e.g. roofs and rainwater gutters) has been investigated (Tilborg *et al.*, 2002 and Gouman, 2002) and the best control at source options appear to be coating the metallic materials with plastic or other materials that prevent exposure and leaching of metals to the environment (Smulders, 2002).

Mohaupt *et al.* (2000) also note that elevated quantities of cadmium in soils (associated with the natural content of the mineral phosphorus deposits prior to extraction for use as fertilizer) are causing concern and the long-term solution for more sustainable agriculture is to use fertilizer with lower cadmium content. Moves to favour the recovery of phosphorus from sewage sludge and farm wastes could help reduce this problem, as well as be more sustainable than quarrying indigenous non-renewable sources.(e.g. Wise and Hann, 2000).

8.4 SUSTAINABLE AGRICULTURE AND FORESTRY

The whole issue of sustainability with regard to agriculture is a vast field and this discussion seeks only to highlight those aspects that relate to diffuse pollution problems and solutions. Broader reviews are presented in OECD (1994) and HELCOM (1998). At the heart of the sustainability concept is the acceptance of human, social needs, including economic as well as environmental considerations. Agricultural and forestry support systems that lead to degradation of vegetation and soil, or divert funds to support practices that have significant off-farm disbenefits and externalized costs, are not sustainable. The Food and Agriculture Organization (FAO) of the United Nations, in *Strategy on Water for Sustainable Agricultural Development* (FAO, 1990) defines sustainable agricultural development as:

> Sustainable development is the management and conservation of the natural resource base and the orientation of technological and institutional change in such a manner as to ensure the attainment and continued satisfaction of human needs for the present and future generations. Such sustainable development in the agriculture, forestry and fisheries sectors conserves land, water, plant and genetic resources, is environmentally non-degrading, technically appropriate, economically viable and socially acceptable.
>
> (FAO, quoted in Ongley,1996).

A classic, albeit apparently rather extreme example of unsustainable agriculture was the degradation of farmland in the former prairie lands of the USA that led to the 'dust bowl' conditions of the 1930s. Less well known but as dramatic, has been the impact of agriculture and deforestation in Iceland (see Text Box 8.3). In the latter example, the Icelandic government Soil Conservation Service (SCS) recognizes that public support for changes in practices is essential. The role of the Icelandic SCS has increasingly been to provide leadership as well as guidance, to all who can affect the condition of the land. Community support programmes are being developed and public participation projects organized, including the involvement of schoolchildren in re-vegetating the land. The SCS 'Soil Conservation 2010' strategy also aims at a shift in philosophy within Icelandic agriculture, towards holistic ecosystem management including multiple uses of land. These shifts in approach seek to realize a broader range of benefits to farmers, as land managers and custodians of the rural environment. It remains to be seen how practical and successful they will be.

Novotny (1999) in a global overview of diffuse pollution and agriculture, considers root causes as well as environmental effects; the inter-related issues of economic and social drivers already explored in some detail in Chapter 7. Success in control is most likely when all three drivers are in alignment.

The Environment Agency/Scottish Environment Protection Agency publication in the UK, *Understanding Rural Land Use* (1999), also highlights socio-economic as well as environmental benefits that are integral to taking steps towards the sustainable management of rural land, arguing that such steps may help to:

Text Box 8.3 Iceland – a case study

The irreversibility or otherwise of the consequences of unsustainable land management is being tested in Iceland. When the first Viking settlers arrived from Norway in the ninth century, they found a green landscape, including woodlands with trees up to 10m tall covering large areas of the country. The settlers brought with them cattle, sheep, goats, and ponies (fundamental to their economy and way of life) – to a country with no native herbivores nor agriculture. Icelandic soils, however, are very susceptible to soil erosion once the covering vegetation has been removed. As populations of settlers and their livestock increased, so the vegetation was over-grazed and woodlands were felled. Severe soil erosion problems ensued and by the late nineteenth and first decades of the twentieth century rates of soil erosion were probably at a maximum. Large districts were devastated by sandstorms, forcing farms to be abandoned. By 1998, vegetative cover was reduced to about 25 per cent of the area of Iceland, with remnants of former woodland down to only 1 per cent. 30,000 to 40,000 km^2 of vegetation lost since island inhabited.

Re-vegetation and restoration

The Icelandic Soil Conservation Service (SCS) was founded in 1907, with the act on 'Forestry and Prevention of Erosion of Land'. The SCS has a role to curtail the soil erosion problem and, latterly, to halt the ongoing destruction of vegetation and soil, reclaim denuded or damaged areas, supervise grazing areas and prevent excessive utilization, protect and improve the remaining vegetation. But the scale of the challenge is daunting: desertification is still affecting 40 per cent of the country. When soils have lost most of their finer materials, soil fertility and water retention are diminished; vegetation becomes vulnerable to disturbance, exposing the soil to wind and water. Restoration methods include fencing off degraded lands to prevent grazing, careful fertilizer application and reseeding using native or introduced grass species, legumes, willows and birch trees. Although restoring at tenfold the rate of loss, it is a long-term project (Runolfsson, 1998; Thorarinsson, 1974).

- reduce the reliance on agricultural production subsidies
- increase the potential income from agri-environment schemes
- reduce inputs, save money and reduce waste
- improve food quality and marketing opportunities reduce risk of pollution
- improve landscape quality and increase the capital value of land and the income from tourism
- improve habitat quality and help to generate income from shooting and fishing
- improve the relationship between the people who manage rural land and the people who wish to enjoy it
- reduce any risk to health and improve the health of livestock and people
- increase biodiversity
- improve water quality and reduce the costs of water treatment
- improve soil stability, structure and condition
- reduce the risk of pests and diseases and the need to use pesticides
- reduce run-off and increase aquifer recharge and dry-season flows.

To favour greater sustainability in agriculture, four leading German scientific associations in the fields of agriculture and water resources management jointly adopted a position paper in 1995 entitled *Agriculture and Environment: An Issue in Need of Political Initiative.* The still up-to-date (see Mohaupt *et al.,* 2000) demands made for the reduction in nutrient inputs are as follows (explanation by Feldwisch and Frede, 1995):

- reorientation of the European Community's agricultural policy to achieve a reconciliation with environmental policy
- modification of quality criteria for products to prevent excessive nutrient use
- redesigning compensation and promotional schemes on the basis of environmentally relevant criteria
- increased integration of environmental aspects in training and counselling
- increased promotion of co-operation schemes between farmers, water companies and administration
- reduction of nutrient emissions from animal production, by
    - limiting the use of farm manure to not more than 40 kg P/ha per annum, additionally to the limits for nitrogen
    - measures to reduce nutrient losses during spreading and storage of farm manure to less than 30 per cent
    - levying a charge on imported animal feed;
- reduction of nutrient emissions from plant production, by
    - fertilization in keeping with the plants' requirements, limiting the nutrient surplus of crop production balances to 50 kg N/ha per annum and 5 kg P/ha per annum

- erosion control measures and set-aside of buffer strips alongside waterbodies
- levying a charge on mineral fertilizers, because the Fertilizers Ordinance did not fulfil all demands
- convening an advisory board dealing with 'agriculture and water pollution control' at the Federal Ministry of the Environment.

The search for more sustainable agricultural and forestry practices requires greater efforts to pull together traditionally disparate strands of government, environmental agencies and others involved in the rural economy and environment. For example, D'Arcy and Frost (2001) argued that landscape measures to control diffuse pollution (buffer strips, for example, but also any of the other measures described in Chapters 2 and 3) are more likely to be established by farmers if several advantages of such action can be weighed against possible disadvantages, rather than simply making a case on grounds of possible diffuse pollution control benefits. This concept of a holistic approach to staking the arguments in favour of environmental protection measures is illustrated in Figure 8.5.

A UK sustainability indicator is bird populations on farms (Figure 8.6). Unfortunately, the data, whilst indicating populations of most UK birds are maintaining themselves, shows the decline of farmland birds in the UK. That decline

Benefits		**Disbenefits**
Cost effectiveness		
Individual benefits for landowner		
Biodiversity		
Amenity	versus	Changes to normal practices
Water quantity		Maintenance
Water quality		Establishment costs

Figure 8.5 Holistic decision-making for land uses considering BMP facilities – tilting the balance in favour of environmental protection and enhancement (after D'Arcy and Frost. 2001). A single, narrow focus on the primary purpose for installing a BMP feature may be insufficient to overcome barriers. When all possible benefits are taken together, BMP's can become cost effective land-use features, with a role to play in rural and urban pollution control.

is associated with intensification of agriculture, including loss of marginal land, pesticide use and other habitat changes: the same changes in land-use practices that cause diffuse pollution. Pollution prevention measures can also benefit the wild bird populations (as well as other species of course). In practice, that might mean that additional habitat or enhanced habitat for farmland birds may be created by the same measures advocated as best practice in appropriate circumstances for pollution prevention. Examples would be the creation of buffer strips alongside watercourses that would also be habitat for small mammal populations on which owls and other birds of prey are dependant, or not ploughing and sowing arable fields in autumn, but reverting to traditional spring ploughing – a practice that would not only minimize soil (and hence phosphorus) losses, but also provide a supply of spilled grain amongst the stubble over winter, for farmland birds such as finches and geese. This is a headline indicator of sustainable development for the UK.

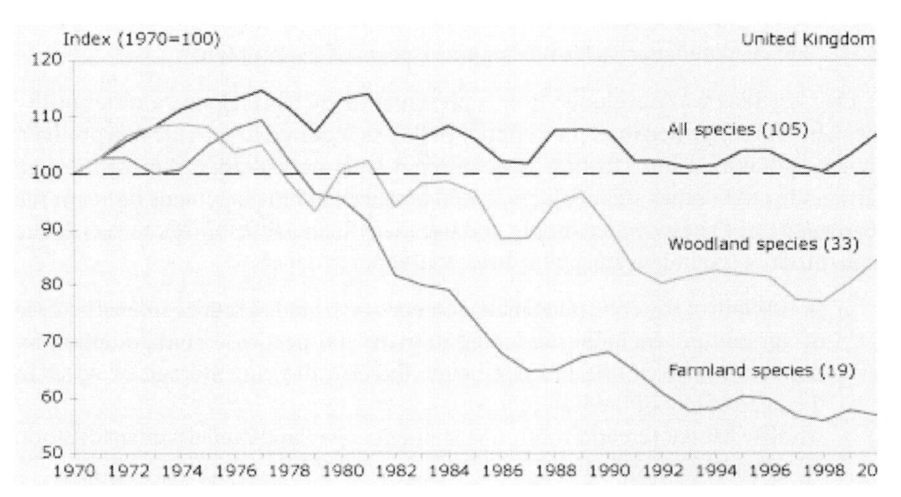

Figure 8.6The decline in the population of farmland birds that has occurred in the UK in recent years 1970-2000 (from DEFRA, 2002). Wild bird populations are considered to be a good indicator of the broad state of the wildlife and countryside. The overall index of populations of British breeding birds has been relatively stable over the last two decades. In 2000, the index of all native species increased by 3 per cent, to its highest level since 1990. But the farmland index almost halved between 1977 and 1993. It decreased by 2 per cent in 2000 after a 3 per cent increase in 1999 and to slightly below the level of 1993. The woodland bird index fell by around 30 per cent between 1974 and 1998 but increased by 4 per cent in 1999 and 5 per cent in 2000 to its highest level since 1990. Although populations of the more common farmland and woodland birds have been declining, rare bird populations, which are not included in this index, have been stable or rising. This reflects conservation efforts focused on these species.

In a remarkable development in 2001, the UK's largest nature conservation organization (the Royal Society for the Protection of Birds, RSPB) teamed with Water UK (an umbrella group for the water utilities) to organize a workshop entitled: *Diffuse Pollution, Floods and Farming* (the full set of presentations plus discussions and actions is available from Water UK, Information & Learning, www.water.org.uk). The workshop involved a search for common ground in dealing with environmental issues across the sectors involved. An integrated perspective on environmental problems associated with agricultural production was presented.

The currently unsustainable features of agricultural production considered by the workshop included the following appraisal of taxpayer/water utility customer perspectives, whereby the public pays three times (Spray, 2002):

1 creation of the problem (production subsidy)
2 the impacts of the problem (externalized costs of agricultural pollution, see Pretty *et al.*, 2000)
3 'end-of-pipe' approach to fixing the aspects of the problem

One key idea was developed from a presentation by Chris Spray, a water utility speaker who suggested that, for a similar outlay as required to provide one product, namely potable water (from a conventional treatment plant), the utilities, in partnership with other stakeholders, could acquire land management rights in the headwaters of the raw water supply and use their financial resources to secure the same product (potable water) plus three additional products:

1 flood control (by controlled management of the land to reduce intensification of agriculture, including reducing nutrient and pesticide contamination by removing or disabling field drainage, thereby allowing storage of water in the headwaters of the watercourse);
2 wildlife habitat (creation and enhancement of wetlands and de-intensification of agricultural production);
3 reduced ecological impacts of diffuse pollution in the watercourse (Spray, 2002).

A fourth product of course is continuing employment for agricultural workers in the headwaters of the water supply river (they would still be needed to manage the land, albeit with a somewhat different remit and additional duties to agricultural production).

That idea was inspired by the actual implementation of such an approach by the cities of New York and Boston, where the water supply has been secured by obtaining land management rights in the headwaters (e.g, in the Catskill mountains), rather than by installing or upgrading a conventional treatment plant – see Text Box 7.8).

There are additional practical opportunities for such 'joined-up thinking' to become normal practice, for example in the EU by the cross-compliance mechanism discussed briefly in Chapter 7. A pre-requisite for arable farmers to receive EU Arable Area Payments for their crops, has been a requirement to take some of the farmland out of production. The cross-compliance element is therefore to put a percentage of the UK arable farmland into 'set-aside' (the purpose being to reduce production). That often means putting some of the farmland into fallow, non-cropped or grazed land uses. When first introduced the required land-take was 17 per cemt, but subsequently that value has been around 5–10 per cent. Real environmental benefits could accrue if that cross-compliance element of the AAP scheme could be modified to include a *requirement* to lay the set-aside land out alongside watercourses as buffer strips, as a first call on the area of set-aside land. Joining up policy and maximizing environmental and economic benefits associated with farming is at the heart of ground-breaking (for the UK) report for the Scottish Executive published in 2002: *Custodians of Change.* The report includes a recommendation that prescriptions for set-aside are revised, on a regional basis, to maximize the accrued environmental benefit and link more appropriately with other government schemes for habitat creation (including riparian zones).

The earlier chapters of this book aimed to introduce information about diffuse pollution as it affects aquatic resources, and details of the broad range of specific measures (BMPs) to address the various facets of the problem. Yet in terms of sustainability and simplicity of control, for livestock farming, there is one single factor at the root of problems and potentially the key to control: livestock density (Table 8.1).

It may be more cost effective to support agricultural activities involving reduced livestock densities, than to finance suites of pollution control measures. Taxpayer subsidies for over-production result in commodities in excess of market demand, and therefore reduce prices and farmer income. Any headage payments in agricultural support schemes should have upper limits, exceeding which incurs penalties such as withdrawal of support.

Such limits have been developed as part of various agri-environment schemes, but often with very limited funds, and applied as supplementary to other mainstream support schemes (e.g. the Rural Stewardship Scheme, as operated in Scotland until 2000, with the aim of encouraging wildlife habitat). The 2003 reforms of the EU Common Agricultural Policy should result in a substantial move away from production subsidies towards economic support for rural environments.

Table 8.1 High livestock density and diffuse pollution.

Problem	*Environmental consequences*	*Mitigation measures*
a) Conventional scenario		
High livestock density	Soil erosion	Buffer strips. Fenced off water margins. Feed points in centre of field. Keep livestock indoors.
	Nutrient enrichment	Transport of manure to other catchments Storage and disposal to land in accordance with farm nutrient budget.
	Faecal pathogens	Slurry digestion and/or high-temperature manure composting Batch storage to avoid application of fresh material to land Buffer zones
	Agrochemicals	Constraints on disposal of manure or slurry contaminated by artificially elevated levels of growth promoters Offsite disposal of sheep dip Collection of drips from dipped sheep on pasture prior to release to farmland and stream crossings
Associated farm husbandry issues control	Greater risks of infection and disease increased use of chemicals increased probability of injuries due to conflict between individuals	Increased costs for farmer animals
b) Alternative		
Lower livestock density	Minimised pollution	Optimised chance for success of mitigation

N.B. without (b), none of the measures in (a) may be sufficiently effective.

8.5 TOWARDS MORE SUSTAINABILITY IN URBAN DRAINAGE AND DEVELOPMENT

8.5.1 An integrated sustainable approach to urban surface water drainage

Run-off treatment and drainage facilities can be designed, constructed and operated as single-purpose facilities, which has often been the practice in the past, or they can be integrated into development plans in such a way that they serve multiple purposes. Treatment facilities can be designed as water features in the urban landscape, lending a sense of open space to an otherwise crowded development.

The term sustainable urban drainage systems as used in the UK and elsewhere in recent years (see Chapter 3) was inspired by the drainage practices developed in the Swedish city of Malmö, and some of the developments in Orlando, Florida (see papers in proceedings of ASCE conference in Malmö in 1997 in Rowney *et al.*, 1999). In Malmö, surface water run-off is not seen as a drainage problem, but as an opportunity to restore wetland areas and provide public open space for informal recreation and to create wildlife habitat (Larsson and Karppa, 1999).

Some 150km to the north of Malmö, but still in the far south of Sweden, is the town of Halmstad, where BMPs have been provided to prevent local flooding in the town and also to protect receiving waters from pollution by run-off. Chief Technical Officer for Halmstad, Torsten Rosenquist, talking about the stormwater ponds there, stated in the original IAWQ video *Nature's Way* that:

> People tend to appreciate them very much and those wetlands have become a target point for walkabouts on Sundays. You walk here, (people) feed the birds, and they stop, they're talking with neighbours here, and there's a social life developing around the wetlands.

In Orlando the use of BMPs to protect groundwater resources from persistent pollutants present in urban run-off, led to the creation of new open space amenities such as the Greenwood Urban Wetlands (case study in Livingston, 1995). 'Lake Greenwood' is a multiple-use facility for flood control, water re-use (irrigation), open space, public recreation, and stormwater treatment prior to discharge down drainage wells (the latter are part of an older system provided to prevent surface flooding and allow aquifer recharge).

Landscape areas and footpaths can be developed in the flood plain, providing the developer with more land. But to be successful, a master plan must be drafted which has the support of all affected parties, *i.e.* the developer, plus municipal planning, highways, and drainage staff, the water and environment authorities/agencies, and the public. A regional-scale case study in the UK is described in McKissock *et al.*, 2001. If the project does not seek and gain this consensus, any one of the parties can cause the integrated plan to fail.

For many larger urban developments, the attenuation of surface water run-off within the development may be essential to prevent flooding. Water quality treatment considerations can be integrated into flood prevention facilities, and *vice versa* if both aspects are properly considered at the outset. Peter Stahre (Malmö City Engineer, pers. com.) believes that

> A key point is the importance of close co-operation among different technical departments in the city. For 'ecological stormwater facilities' in the urban environment the challenge is not so much a technical one as an institutional one. Basically we have a very good knowledge of how to calculate, design and construct this type of facility. That is not the problem. What we must learn is the strategy to sell the idea to the city administration. The key to success is multiple use and interdepartmental co-operation.

8.5.2 The unsustainable aspects of urban drainage systems

Conventional drainage systems concentrate pollutants in urban run-off and allow them to be drained swiftly and directly into the nearest watercourse. Research undertaken by the Scottish Environment Protection Agency, SEPA, (Wilson and Clarke, 2002) involved a survey of nine urban streams in Scotland and found high levels of persistent pollutants in samples of sediment from the stream beds (Tables 8.2 and 8.3). In four of the nine streams, if the sediments were to be dredged they would be classed as 'special waste' – and many of the sites would yield material unsuitable for application to farmland or for use in urban landscaping. The relative abundance of the toxic metals present in the sediment samples (although showing considerable variation between sites) suggested road traffic was a significant source.

The notable presence of 4–6-ring PAH compounds in a high proportion of the sediment samples also indicated road traffic (internal combustion engine) sources as generally more important than for example, oil spills; although again there was variation and at a few of the industrial sites a prevalence of 2–3-ring PAHs indicated that oil spills were a dominant influence (Wilson and Clarke, 2002).

The ability of BMPs to trap and retain persistent pollutants does not make the use of BMPs (or SUDS – see Chapter 3) unsustainable; the latter term should instead be applied to the processes and activities that use persistent pollutants and introduce them into the environment.

A clearer focusing of attention on problems at source (use of toxic metals, and traffic management and provision of alternative public transport options) would leave engineers and landscape architects with stormwater management options as described in Chapters 3 and 4 that can deal effectively with unavoidable residual levels of contamination, particularly by degradable pollutants, nutrients and sediment. In the interim the retention of persistent pollutants in BMPs is preferable to dispersal into urban stream sediments and ultimately coastal environments (Wilson

Table 8.2 Sediment quality: Ontario provincial sediment quality guidelines, (Ontario Ministry of the Environment, 1993) and threshold values for UK Special Waste Regulations, 1996 (from Environment Agency, 2002b)

a) Ontario provincial sediment quality guidelines (values quoted in mg/kg dry weight unless stated)

Determinand	*Lowest Effect Level*	*Severe Effect Level*
Arsenic	6	33
Cadmium	0.6	10
Chromium	26	110
Copper	16	110
Lead	31	250
Nickel	16	75
Zinc	120	820
Total organic carbon (%)	1	10

b) UK special waste regulations

Determinand	*Total Concentrations*		
	Leachate quality threshold	*Lower threshold concentration (mg/kg air-dried sample)*	*Upper threshold concentration (mg/kg air-dried sample)*
PH	5.5–9.5 μg/l	6–8	5–9
Toluene extract	–	5000*	10000*
Cyclohexane extract	–	2000*	5000*
Conductivity	1000 us/cm	–	–
COD	30 mg/l	–	–
Ammonia	0.5 mg/l	–	–
Arsenic	10 μg/l	10	40
Cadmium	1 μg/l	3	15
Chromium (total)	50 μg/l	600	1000
Lead (total)	50 μg/l	500	2000
Mercury	1 μg/l	1	20
Selenium	10 μg/l	3	6
Boron	2000 μg/l	3	–
Copper	20 μg/l	130	–
Nickel	50 μg/l	70	–
Zinc	500 μg/l	300	–
Cyanide (complex)	–	250	250
Cyanide (free)	50 μg/l	25	25
Sulphate (SO_4)	150 mg/l	2000	2000
Sulphide	150 mg/l	250	250
Sulphur (free)	150 mg/l	5000	5000
Phenol	0.5 μg/l	5	5
Iron	100 μg/l	–	–
Chloride	200 mg/l	–	–
Polyaromatic hydrocarbons	0.2 μg/l	50	1000

Note * Subject to special waste

et al., 2003). Maintenance regimes (removal of contaminated sediments) need to be designed to minimize bioaccumulation or other toxicity risks.

The other aspect of conventional drainage networks that is arguably less than sustainable is the increased frequency and significance of local flooding in small urban watercourses as a result of urbanization. This consequence of positively drained imperviousness is another reason why urban engineers often spend the environmental capital (as well as community funds) of their towns; putting streams and rivers into concrete culverts, where no wildlife can survive and there can be no public benefit from having a natural watercourse for informal recreation. Again, properly designed BMP facilities can be developed to mitigate these impacts and reduce the need for streams to have hard engineering reinforcement of bed and river bank.

There is a sustainability issue, too, in the choice of options to manage stormwater in older parts of many cities, where a single foul sewer network receives all inflows (sewage and run-off in a combined system). The use of distributed storage, targeting the largest areas of imperviousness in the sewer catchments such as car parks at shopping malls, pedestrianized town centres, and school playgrounds, would allow utilities to put the charges and preventative actions at the sources of the stormwater management problems.

In some situations that might mean conventional below-ground storage tanks, oversized pipes, or vortex-flow control devices. In others (e.g. school playgrounds)

Table 8.3 Scottish Urban streams sediments survey, 2001. Sites exceeding guideline standards (c.f. Table 8.2). (After Wilson and Clarke, 2002, unpublished report of Scottish Environment Protection Agency, SEPA).

Site	*Determinand*	*Lower threshold concentration (mg/kg)*	*Standard*	*Value (mg/kg)*
East Tullos 2	Copper	130	ICRCL	440.6
	Nickel	70	ICRCL	80.9
	Zinc	300	ICRCL	407.0
Kittoch Water 3	Nickel	70	ICRCL	192.8
Lyne Burn 1	Zinc	300	ICRCL	411.7
Findon Burn 2	Zinc	300	ICRCL	303.0
Caw Burn 3	Arsenic	10*	ICRCL	17.65
Town Loch 1	Σ•PAH	50	ICRCL	52.5
Caw Burn 1	Total hydrocarbons	1000	Special Waste Regs (1996)	3382
Dedridge Burn 1	Total hydrocarbons	1000	Special Waste Regs (1996)	2175
East Tullos 2	Total hydrocarbons	1000	Special Waste Regs (1996)	1641
Lyne Burn 1	Total hydrocarbons	1000	Special Waste Regs (1996)	1603

Note
* Threshold set regarding domestic gardens and allotments.

more creative semi-natural wetland features could achieve the same flow control objective and provide an educational feature for the schoolchildren. The conventional approach on combined sewer networks has been to gain access to large plots of land at public expense and dig enormous storm sewage tanks, complete often with banks of stormwater pumps and sophisticated self-cleaning screens to treat storm sewage overflows to watercourse. The economic costs as well as ongoing operational costs and energy consumption, need to be compared with serious efforts to consider alternative source control options (see Yorkshire, UK case study for example Stovin and Swan, 2002). Those conventional costs are the real price to pay for total intolerance of occasional surface flooding within the urban environment.

Finally, reference is made in the following section to protecting the self-purification capacity of urban watercourses by retaining meanders, riffles, riparian wetlands and flood plains from inappropriate development: also important in the towns and cities of the developed world.

8.5.3 Breaking the pollution cycle in developing cities

The approaches to deal with urban run-off outlined in Chapters 2 and 3 and in particular the opportunities to retrofit solutions as urban areas redevelop, noted in Chapter 4, are practical options in developed countries with adequate resources to fund redevelopment, albeit at a slower rate than is desirable. But in the growing urban areas of the developing world the pollution problems seem daunting.

In South Africa, diffuse pollution has been identified as the major threat to water resources, with a legacy of problems associated with mining, burgeoning problems with new urban and peri-urban populations, and conflicting demands on water resources from agriculture, nature conservation and tourism, industry, and domestic potable supply utilities (see Table 8.4, after Heath, 2000). The impacts on water quality suggest that it will be even more difficult for the public water supply utilities to meet the rising demands for clean water. Some of the most severe problems are from unplanned settlements in the fringes of existing cities, resulting in heavily polluted run-off and contamination of groundwater. The traditional approaches of trying to extend conventional sewerage networks to pick up many of the diffuse sources are hugely expensive, and often only partly deal with the problem. In several parts of Africa a variety of approaches are being developed, looking in an integrated way at pollution, at consequent health issues, and at the need for safe potable supplies. The need for local community buy-in is paramount.

In South Africa, the Department of Water Affairs and Forestry (DWAF) has produced a *Framework for Implementing Non-Point Source Management*, that acknowledges the importance of community involvement in controlling diffuse pollution (DWAF, 1998). The principle of public involvement, and ownership of water resource management by the communities, is the basis of the South African National

Table 8.4: Water quality issues as a result of diffuse pollution from an African megacity (Heath, 2000)

Source of diffuse pollution	*Resultant problems*
Urban runoff after rains	BOD, microbiological, litter, *Giardi*, *Crytosporidium*, cholera, eutrophication
Groundwater contamination	Microbiological, nitrogen, toxins
Blocked and leaking sewers	BOD, microbiological, litter, *Giardi*, *Crytosporidium*, cholera, eutrophication
Atmospheric deposition (factories, cars, burnt fossil fuels)	Sulphur dioxide, nitrogen dioxide, total suspended particulates
No reticulation	BOD, microbiological, litter, *Giardi*, *Crytosporidium*, cholera, eutrophication
Pit latrines on clay soils or high water tables	Surface and ground water contamination
Overloaded waste water works waters	Increased microbial and nutrient loads to surface
Urban subsistence agriculture	Soil erosion, habitat loss, fertilizers (nutrients), biocides, faecal loads (sheep, cattle, pigs, chickens), *Giardi*, *Crytosporidium*
Storm water draining into waste works	Overloading and inefficient treatment
Mine dumps	pH, salinity, sedimentation, manganese, aluminium, sulphates, arsenic, toxicity
Mine water discharged on surface	pH, salinity, sedimentation, manganese, aluminium, sulphates, arsenic, toxicity
Increase runoff due to increased impermeable surfaces	Flooding, waste works overloaded
Habitat destruction (riparian zone)	Safety hazards when floods, erosion, loss of habitat
No refuse removal	Litter, contaminated water

Water Act (Act 36 of 1998). Rivers forums enable the communities to get together to learn, discuss and ultimately implement grassroots catchment management. Environmental education, as discussed in the previous chapter), is critical for the sustainable understanding, ownership and management of the water resource. An excellent DWAF public education booklet describes the relationships between poverty, pollution and health in a concept it calls the Pollution Cycle, whereby:

Inadequate services are followed by poor health, a polluted environment, poverty arising from inability to work, and hence low payment of service charges - which further leads to inadequate service provision.

A downward spiral of problems as illustrated by the South Africans, is broadly applicable in many other countries with burgeoning settlements associated with megacities. A slight modification of the DWAF figure, as below, suggests opportunities for breaking the cycle, utilizing simple best-practice technology allied to a low-cost, low-maintenance approach. Community acceptance of measures, with direct involvement of community leaders in development of options, is a pre-requisite for success.

A first measure to break the link between the polluted environment and the health of livestock and the human population, could be the provision of riparian buffer zones, as in Figure 8.7 based on considerations in Table 8.5.

The difficult issues of river water quality and public health could be further addressed by avoiding the temptation to try and achieve all water quality targets at once. Iwugo *et al.* (2003) have suggested that a phased approach to pollution control

Table 8.5: Problems and possible partial solutions for riparian zones

Pollution issues with riparian zones in peri-urban communities
• Easy access to polluted water for potable use • Traditional use of streams for polluting activities (e.g. laundry, bathing) • Easy stream access for livestock, with consequent health risks for livestock and human population (as above, often no alternative, of course) • Ad hoc refuse dumps for settlements often on edges of watercourses, contributing to health risks and pollution • Loss of self-purification capacity of watercourse as settlements encroach on streams with infill of meanders and natural wetlands
Urban buffer zones: simple fencing, riparian hedgerows and wetlands to:
• Prevent access for livestock to polluted water supplies (water troughs provided as alternatives) • Discourage but not preclude, access for people (easier to use safer water supplies) • Reduce refuse impacts: if a fence/ hedge is far enough back from the river, even if some people continue to dump wastes over the barrier, refuse does not get into the river • Establish alongside rivers, vegetation that should help stabilise banks and filter some of the surface flows of polluted runoff • Protect natural self purification capacity in the streams by protecting natural riparian marshes, pools and meanders from infill by refuse deposition or development • Provide and enhance, for the local community, wildlife and amenity interest in their watercourses.

(a) Without interventions

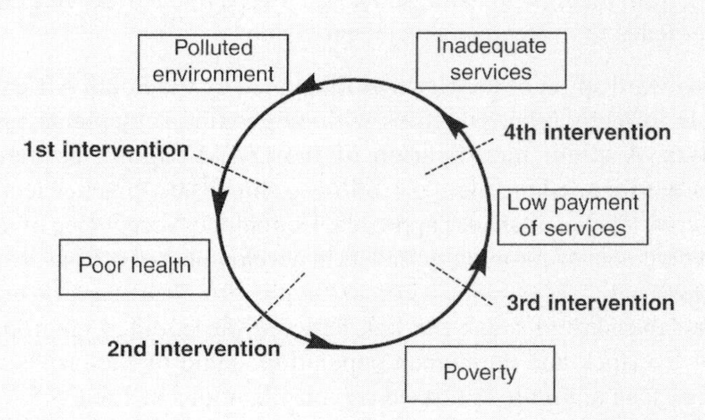

(b) With appropriate interventions - sustainable environmental technologies and socio-economic measures.

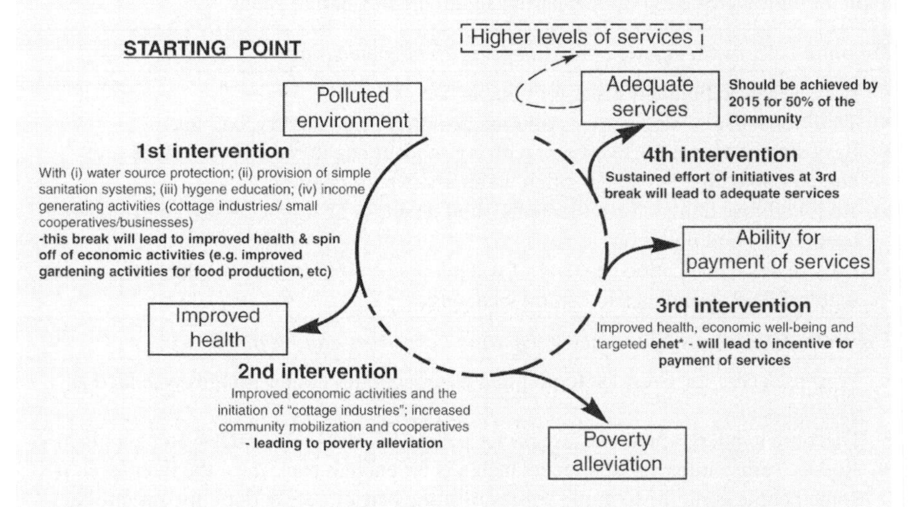

*The problems could be addressed by **breaking the links** in the cycle, starting with the link between polluted environment and health of people and livestock. [e.g. discourage access to unsafe water (rivers and streams) whilst providing alternative supplies].*

*environmental health education and training

Figure 8.7 Breaking the Pollution Cycle (I) (anti-clockwise cycle in figure below) (Iwugo *et al.*, 2003)

is unavoidable for new and existing unplanned urban settlements where socio-economic instability and the pace of change requires a departure from conventional strategies for sewerage and treatment infrastructure.

Tables 8.6 and 8.7 indicate problems and possible options. Provision of low-capital-cost, low-maintenance treatment and drainage schemes may not achieve any improvement in the nutrient status of receiving waters, but technology such as septic tank and reedbed systems in combination, offers excellent removal of oxygen demand, faecal pathogens, and suspended sediment, with significant removal of ammonia (as advocated in Iwugo *et al.,* 2001). Passive treatment systems for stormwater will also reduce diffuse loadings. This suggests a further approach to breaking the Pollution Cycle, as indicated in Figure 8.7 (Iwugo *et al.*, 2003).

Table 8.6 Breaking the Pollution Cycle (II): Matching socio-economic conditions to environmental improvements

Social conditions	Implications	*Possible water service provision*
Inability to pay Unwillingness to pay	Low capital cost, and low operating costs.	Stand-pipes for potable water. Local treatment, e.g. septic tanks and reedbeds (not trunk sewers with pumping stations and sophisticated treatment plants)
Hostility to outsiders/ national agencies	Low maintenance requirement.	
Community dependent upon 'black market' income.	Drainage facilities should not involve components of significant economic or functional value.	Local treatment as above, passive treatment systems, sub-surface treatment (e.g. septic tanks and reedbeds).
Strong self-help culture and desire for environmental and economic improvements.	Engagement with local community by public agencies/services. Community support reduces theft and vandalism.	Dialogue on needs and programme of improvements.

Table 8.7 Breaking the Pollution Cycle (III): Interim, low-cost capital projects for poorly serviced settlements

- Grass swales draining into fully vegetated detention areas prior to discharge to river are better able to attenuate and degrade contamination arising from grey water disposal, leaks from chemical toilets etc.
- Fully vegetated detention areas on surface drains intercept polluted sediment in runoff, and allow some biodegradation between storm events.
- For foul flows, septic tank followed by root-zone reedbed treatment offers low maintenance, safe and inaccessible passive treatment technology.
- The operational life of such passive treatment sytems may be consistent with the anticipated lifespan of some settlements.
- Septic tank and reedbed systems offer very limited opportunities for scavenging any materials of value for black-market use elsewhere (no pumps, metal components of scrap value, etc).
- Septic tank and reedbed (root zone) systems both offer minimal risks to health and safety for the local community who may gain entry to premises (passive processes, below ground surface).
- In combination, pollutant removal efficiency for first priority contaminants (primarily faecal pathogens and gross solids, but also oxygen demand) is high.

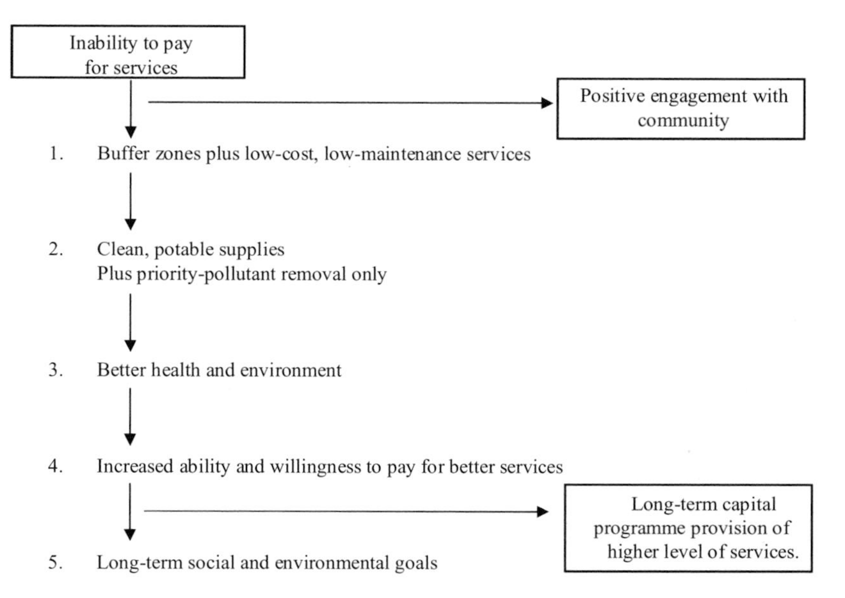

Figure 8.8 Relating socio-economic condition to progressive environmental improvements

8.6 CATCHMENT MANAGEMENT AND SOCIAL ISSUES

8.6.1 Introduction

All the above sections included reference to social dimensions of effecting changes in practices, whether in agriculture, urban development, or the apparently intractable problems of megacities. Community buy-in and public support are pre-requisites for the success of any programmes. An effective way of securing that buy-in is to show the relevance of issues to people in a local context. A river catchment is a natural geographic entity that is readily recognized by populations within it. There is also scope for narrower local initiatives, just as there is for regional, national and international mobilization of concerns and pursuit of sustainable solutions. The remaining sections of this chapter are concerned with the wider consideration of public interest in, and demand for, environmental protection and improvement.

8.6.2 Catchment management

The Fraser River Basin, in British Columbia, Canada, drains a hinterland of alpine uplands in the Rocky Mountains, through forests and farmland to the city of Vancouver. The Fraser River is one of the greatest salmon providing river systems in the world and the river basin has enormous wildlife and recreational value. Yet the basin supports two-thirds of the province's population, and its business activity accounts for 80 per cent of the province's gross domestic product, and 10 per cent of Canada's gross national product. The river basin is at a turning point (Marshall, 2002), facing pressure from population growth, resource extraction and development. In the next 25 years overall population is expected to increase by 50per cent, and in some parts of the basin by 200 per cent. In recognition of the growing need to secure long-term sustainability, the Fraser Basin Council, a not-for-profit society, was established in 1997. It has become an unprecedented governance structure, in which the four orders of Canadian government (federal, provincial, local and indigenous people), the private sector and civil society work together to solve many complex inter-jurisdictional issues affecting the sustainability of the watershed. The environmental issues addressed include diffuse pollution associated with forestry, livestock and arable farming, mineral extraction, and urban run-off (a significant issue in small tributaries in and around Vancouver, for example) (Marshall, 2002).

The Fraser River basin management forum described above is similar to the approach for more engagement with local communities and interests, advocated in Europe via the EU Water Framework Directive (see website http://europa.eu.int/water/water-framework). It also parallels the consultative management process established in South Africa under their National Water Act (Act 36 of 1998), which endorses the principles of integrated water resource management. The latter is driven by a deep-seated public consultation and education process (Heath *et al.,* 2002). The process of implementation is being piloted in three of the nineteen new nationally designated river catchments (as of September 2002).

One of the original drivers for BMPs to control diffuse pollution in the USA was concern for water quality in coastal zones, including Chesapeake Bay. There, an exemplary effort to adopt a more sustainable approach to development and to control of diffuse pollution has been mounted in recent years, in parallel with a smaller, but equally useful, catchment project in Australia (reported as a combined case study in sustainable development and management of pollution by Dennison (2002). The approach classically links clear establishment of the facts driving necessary actions (e.g. the importance of sediment and nutrient loadings associated with agriculture and forestry) to problems evident in the coastal waters, with sector buy-in to the solutions. Public involvement programmes feature in the catchment/ river basin work, alongside research and regulation. The community involvement includes a river festival, for example, in Chesapeake Bay.

A series of small urban streams in two catchments in Scotland illustrate another level of approach to more sustainable management of watercourses and control of diffuse pollution. The Dunfermline East Expansion (DEX) is an area of some 48 km^2, planned to be the site of 5,500 homes as well as business and leisure developments, on the upstream side of the old town of Dunfermline (a medieval seat of Scottish kings). The urbanization of the greenfield land in the headwaters of two small streams that pass through old culverts in the town would have resulted in greatly exacerbated flooding problems for the local population. In a hitherto unique UK project, a partnership of regulators, developers and enterprise bodies was formed to oversee the design and implementation of a drainage scheme that would prevent flooding, prevent diffuse pollution problems arising from the development, and create new wetland habitats and amenities for the local population. The development history and the process of establishing an acceptable drainage system is described in D'Arcy and Roesner (1999) and McKissock *et al.,* (2001). More recently a community engagement project has been established to explain the drainage scheme and co-develop amenity and habitat opportunities with the local community (adjoining existing population as well as new home owners at DEX). The latter involves the regulator (SEPA, Scottish Environment protection Agency), the local authority (Fife council), the water utility (Scottish Water) and a range of commercial developers. A second project has been established for a major new development and re-development in and close to Edinburgh, led on behalf of a similar partnership of regulators and local authorities, by SISTech, Heriot-Watt University (see Text Box 8.4).

Text Box 8.4 The Edinburgh South East Wedge development: Sustainable Urban Drainage Systems – Maximizing the benefits through integrated planning and working in partnership

A unique partnership group has been established to promote the sustainable management of surface water in a major development known as the – South East Wedge – on the outskirts of Edinburgh (Scotland). This group comprises local planning authorities, environmental regulators, engineering consultants, NGOs, and the water utility. The aim of the group is to ensure that the environmental potential of the proposed development is maximized through the integration of amenity, landscape, habitat and urban drainage. Project Management been provided by the Scottish Institute of Sustainable Technology (SISTech).

The 'South East Wedge' development area covers a total area of 1,370 hectares and comprises land both within Midlothian and Edinburgh. Eventually almost 6,000 new homes will be built, together within commercial and other developments.

Three relatively small streams (already impacted by urban run-off, combined sewer overflows and other discharges such as pumped minewater), drain the development area. These streams have also been subjected to historical culverting or channel modification. In some cases culverting has served to increase the flood risk. The integration of SUDS into the development should not only minimize the impact of the development on these burns, in terms of water quality and flooding, but also provide significant ecological, landscape and amenity value. Returning the watercourses to a more natural state through de-culverting and re-introduction of natural watercourse features, provides an opportunity to reintroduce burns as a community focus, and could alleviate flood risk associated with troublesome culverts.

The cornerstone of the group's work is the production of the *Environmental Enhancement Action Plan for the South East Wedge.* This provides a model framework and technical guidance for the integration of SUDS, and local watercourse enhancement, in the development.

For SUDS, the Action Plan focuses on the use of the 'surface water management train' to ensure the protection of receiving watercourses and minimize the potential for increase in flood risk downstream. A typical 'source-site-regional' treatment train could consist of the use of porous paving in areas such as driveways and car parks, followed by detention basin, and finally a retention pond or wetland before discharge to the watercourse. Overall, it is proposed that a total of 17 regional ponds or wetlands would be required.

A central feature of the Action Plan has been the positive integration of SUDS features within the overall landscape and ecology strategy for the development.

Text Box 8.4 continued

If designed appropriately, regional ponds and wetlands can serve as focal point for housing developments as well providing important natural habitat. Linear features, such as swales, can form part of a network of green corridors that provide safe routes for both wildlife and people. The Action Plan serves as a vital point of reference for planners, developers, and engineers who are involved in implementation of the development.

(Tim Darlow, SISTech, Heriot-Watt University, www.sistech.co.uk)

8.6.2 Public expectations

As the quality of life improves, the expectations of the public with regard to their environment increases. The beginning of the last century saw the development of popular 'green' movements in the USA, South Africa and Europe, for example in the USA with the founding of the influential Sierra Club (1892) and Audubon Society (1905). The first national parks in the world were established in the in the USA even before that time, e.g. Yellowstone in 1872, Yosemite and Sequoia in 1890. In South Africa a similar response to increasing wealth and the disappearance of wilderness, as development proceded across the country, led to the foundation of the Kruger National Park in 1898. Those early concerns for nature and wild places became more general concerns for a broader view of the human environment, particularly after the Second World War with the declines in air and water quality, and loss of land to increasingly intensive development in the post-war years. Evidence of that growth in public environmental concern is the number of new non-governmental organizations (NGOs) established during the post-war years, (many with international as well as national interests) together with the growth in support for older organizations. A few examples are the World Wildlife Fund (WWF – now the World Wide Fund for Nature), the Friends of the Earth (FoE), Greenpeace, the Wildfowl and Wetlands Trust; the list keeps growing and in ever more countries, including active organizations in many developing countries.

Various surveys in the developed world have shown that the quality of the physical environment is greatly valued by the public, and in 1993 a report for the Scottish Office (government in Scotland at that time) indicated that water pollution was top of the list of environmental concerns. Successive UK government surveys in England and Wales found similar findings; e.g. in 1996/7, that that the top two pollution concerns were water pollution issues (chemicals put into rivers and sea, and sewage on beaches/ bathing water) and subsequent surveys continue to show the importance of pollution concerns (DEFRA, 2002).

Broad environmental concerns are one end of a spectrum of interest that starts with the basic need for an adequate and wholesome supply of drinking water. This has been recognized for example by the new South Africa, where the then Water Minister in President Nelson Mandela's initial post-apartheid era government, identified a target of 25 litres per head per day of potable water as a basic human right for everyone in South Africa. Even that is an awesome aspiration: in South Africa in 2002 it was estimated that approximately 7 million people do not have access to adequate supplies of potable water, and nearly 18 million people lacked basic sanitation services (DWAF, 2002). Many of the problems of contamination are diffuse in nature (Heath *et al.*, 2002 and Table 8.1 above). Resolution of these pollution problems is not a developed world quality-of-life issue, it's a basic necessity.

Public support for control of environmental degradation and pollution, as well as recognition of the economic importance of protecting water resources, is evident in the development of ever more comprehensive environmental regulation. In the UK, for example, this is evident by the history of water pollution legislation. Initial legislation at the end of the nineteenth century, was followed by a fifty-year gap, then successive new Acts of Parliament, initially at almost ten-year intervals, increasing in 1990s and latterly with European Union legislation too, at ever-decreasing intervals. That history also shows the increasing scope of legal measures to try and control pollution. Often problems associated with a lack of action to fully utilize existing statutory means of persuasion were confused with a need for new legislation. Regulators and regulated sectors are now faced with an unmanageable regulatory mushroom, a situation that is also prevalent in the USA. It is arguably an unsustainable approach to achieving environmental goals. Regulation needs to be aligned with economic drivers as well as social needs and expectations, and amongst the social needs is the need for simple regulatory controls, not a complex plethora of statutes, exemptions and specialist applications (see Chapter 7).

Morgan Williams (2002), in his excellent keynote presentation to an International Water Association sustainability conference in Venice in November 2002, quoted Maurice Strong the former Secretary-General of the 1992 Earth summit addressing a US Senate Environmental Treaty Implementation review in July 2002. Mr Strong said:

> Most of the changes we must make are in our economic life. The system of taxes, subsidies, regulations and policies through which governments motivate the behaviour of individuals and corporations continue to incentivize unsustainable behaviours.

Poul Harremoes, in another valuable keynote presentation at that Venice conference, concluded that sustainability is essentially an ethical, rather than a technical issue. In relation to water issues, it is an intention to provide global equity and secure fundamental rights to water. Decisions with respect to sustainability, environment, water, human health and provision of food should be based on a combination of multi-disciplinary (scientific, technical, social, economic, and ethical) analyses. In view of scientific uncertainty and the difficulty of forecasting future resource availability and environmental conditions, decisions should aim at precautionary prevention and be more adaptive to change (Harremoes 2002).

To those guiding thoughts should be added a maxim for regulators, regulated sectors, and others hoping to achieve more sustainable ways of generating wealth, and protecting health and the environment within the confines of existing statutes, rules and institutional practices and traditions:

Work together, and make it happen!

8.6.3 A cause for optimism?

The examples given in Chapters 4 and 6 of demonstration projects and reviews of the effectiveness of best-practice technologies, suggest that there are cautious grounds for optimism that diffuse pollution can be controlled effectively. But will it be? The examples given earlier in this chapter of restrictions on use of damaging chemicals whose side-effects (environmentally) were unforeseen, show that mistakes can eventually be rectified, but how many more will be made and at what price to the environment? History suggests that screening for environmental impacts of chemicals is inadequate.

The range of persuasion tools available to governments and non-governmental organizations (NGOs), as indicated in Chapter 7, offer a flexible suite of options for effective action. Technical knowledge and experience exists, but is there political will to put in place measures to ensure more sustainable land use practices? The economic cost of inaction, as well as the cost of unsustainable policies such as the much criticized Common Agricultural Policy (CAP) in the EU, for example, also offer grounds for belief that necessary reforms and controls will be eventually effected. But how much money will continue to be spent on unsustainable practices (e.g. subsidies for over-production and associated environmental damage) before the political arguments are resolved?

This concluding chapter has focused on how diffuse pollution of aquatic resources is a sustainability issue; long-term threats to the environment and to the economic well-being of society. The solutions require the integrated responses that are at the heart of sustainable development considerations. Many solutions have the potential to bring, or be part of, broader social and economic benefits, whether through more sustainable agricultural support schemes and rural land-use diversification, through protection of water resources and health, or through restoration of damaged aquatic resources with consequent benefits for productive employment.

In many parts of the world, there is growing popular support for environmental concerns. For example, by 2000, the membership of one UK nature conservation body, the RSPB (Royal Society for the Protection of Birds), exceeded one million – probably more people than the combined membership of all the UK political parties. The role of non-governmental organizations is important: their efforts need to be vociferous, professional and sustained. Ultimately, that sort of popular demand for environmental protection, together with international recognition that water is a fundamental human requirement, suggests that there should be a political will to succeed.

Intensive agriculture, is the main source of excessive nutrient loads into the Venice lagoon

Road traffic and run-off is a major issue for diffuse pollution and for human health, as well as from social and economic considerations

Provision of cycle tracks addresses diffuse pollution sources, as well as aiding efforts to achieve safer and reduced traffic flows

Considerations for urban drainage arising from new developments, need to include flooding risks and the requirements for additional capacity in existing infrastructure unless flows are attenuated by BMPs

Megacities require a strategic approach to pollution control that is entirely dependant on social inclusion and active community involvement (West Africa)

Community engagement involves telling local people what features of their environment are for – e.g. the signs erected at BMP facilities in Halmstad, Sweden

Abandoned mines often generate highly polluting flows from springs as water tables rebound after cessation of pumping (S Africa). Adverse impacts on water resources persist for decades

The public value water as a recreational resource, as well as clean unpolluted water for environmental and economic uses (photos from Orlando, Florida, and school visit to BMP pond in Dunfermline, Scotland UK).

Appendix 1

Effectiveness of Rural BMPs

Extract from J Hilton, *BMPs Dictionary*, Centre for Ecology and Hydrology, Wallingford, UK, 2003.

INTRODUCTION

The table below outlines some figures from the literature on the pollution reduction effectiveness of a range of BMPs. The data is drawn from the *BMPs Dictionary*. A wider set of referenced data is being included in an update of the original dictionary, to produce a BMPs manual in early 2005, that will be available from the Scottish Environment Agency, SEPA. The table below includes the main types of BMPs, but many more are described in the *BMPs Dictionary* and its successor document (in preparation). The original and seminal work descriping BMPs and their costs and effectiveness is the document produced by the US Environmental Protection Agency in 1993: *Guidance Specifying Management Measures For Sources of Nonpoint Pollution in Coastal Waters* and the update information on the USEPA website: http://www.epa.gov/owo/nps/MMGI

The effectiveness values quoted below are inevitably variable: this is to be expected given the range of conditions in which BMPs have been applied and assessed. Misapplied measures can be expected to be ineffective, and local climate and soil conditions, as well as land use and practices will further determine pollution control that can be achieved by the range of possible measures.

BMP description	*Phosphorus*		*Nitrogen*		*E.Coli*		*Suspended solids*		
	max	*min*	*max*	*min*	*max*	*min*	*max*	*min*	*Comment*
Use of soil P analyses to indicate over fertilization	25	0							
Location of the fertilizer closer to the plant root (US = fertilizer banding)	40	0							
Timing of fertilizer applications – Split application of fertilizer/slow release fertilizer	5	0							
Nutrient management plan	90	20	90	20					
Nutrient management plan									
Waste (manure) management	21		62		74		64		
Prevention of misplaced fertilizer (spills)	15	0							
Critical area planting (changing land use)	90	0							
Grassland rotation	50	50	30	30					
Cover crops	50	30					60	40	
Conservation tillage 1 – Crop residue management							75	75	Residue = 0.25 t /acre
Conservation tillage 1 – Crop residue management							92	92	Residue = 0.5 t /acre
Conservation tillage 1 – Crop residue management							97.5	97.5	Residue = 1.0 t /acre
Conservation tillage 2 – Till planting	85	35	80	50			90	30	
Conservation tillage 2 – Till planting							86	86	Maize, 34% cover
Conservation tillage 3 – Strip rotary tillage	85	35	80	50			90	30	
Conservation tillage 3 – Strip rotary tillage							76	76	Maize at 27% cover
Conservation tillage 4 – No-till planting	85	35	80	50			90	30	
Conservation tillage 4 – No-till planting							96	96	Wheat at 86% cover
Conservation tillage 4 – No-till planting							64	64	Soya at 27% cover
Conservation tillage 4 – No-till planting							92	92	Maize at 39% cover
Conservation tillage 5 – Annual ridges	85	35	80	50			90	30	
Conservation tillage 6 – Chisel plough	85	35	80	50			90	30	

BMP description	*Phosphorus*		*Nitrogen*		*E.Coli*		*Suspended solids*		
	max	*min*	*max*	*min*	*max*	*min*	*max*	*min*	*Comment*
Conservation tillage 6 – Chisel plough							72	72	Wheat at 29% cover
Conservation tillage 6 – Chisel plough							32	32	Soya at 7% cover
Conservation tillage 6 – Chisel plough							74	74	Maize at 35% cover
Conservation tillage 7 – Discing	85	35	80	50			90	30	
Conservation tillage 7 – Discing							26	26	Soya at 8% cover
Conservation tillage 7 – Discing							72	72	Maize at 21% cover
Livestock exclusion	90	50					90	50	
Strip cropping	50	50					75		
Strip cropping							7	7	
Contour cropping							25	25	
Hedgerow planting							71		
Terracing (or contour bunds)	66	66					14	14	
Contour cultivations							90	30	
Contour cultivations							50		
Water retention systems – 2 – Grass waterways	45	15	45	15			65	30	
Water retention systems – 2 – Grass waterways							50		
Water retention systems – 2 – Grass waterways								80	
Water retention systems – 3 – Detention basins	77	77	37	37			85	85	
Water retention systems – 3 – Detention basins	45	15	45	15			65	30	
Water retention systems – 3 – Detention basins							74	70	
Water retention systems – 4 – Retention ponds	60	10							
Water retention systems – 4 – Retention ponds	54	54							
Water retention systems – 4 – Retention ponds	79	0	80	0			91	91	

BMP description	*Phosphorus*		*Nitrogen*		*E.Coli*		*Suspended solids*		
	max	*min*	*max*	*min*	*max*	*min*	*max*	*min*	*Comment*
Water retention systems – 4 – Retention ponds	65	30	65	30			80	50	
Artificial reed beds	46	46	24	24			76	76	
Artificial reed beds							90		
Wetland restoration	70	10	80	40			90	80	
Wetland restoration							90		
Oxidation pond sedimentation basin							90		
Oxidation pond sedimentation basin	40	40					87	40	
Riparian buffer zones – 1 – protection from machinery operation	15	0							
Riparian buffer zones – 1 – protection from machinery operation	82	11							5m width
Riparian buffer zones – 2 – solids reduction	80	50	80	50			80	50	
Riparian buffer zones – 2 – solids reduction							90	35	
Riparian buffer zones – 2 – solids reduction								80	5m width
Riparian buffer zones – 2 – solids reduction							82	11	10m width
Riparian buffer zones – 2 – solids reduction							82	13	
Riparian buffer zones – 2 – solids reduction							25	5	
Riparian buffer zones – 3 – dissolved pollutant reduction	75	50	90	80			90	80	

Appendix 2

The International Stormwater BMPs Performance Database

(Contributed by Ben Urbonas, 1994, Denver Urban Drainage and Flood Control District)

The Urban Water Resources Research Council of the American Society of Civil Engineers developed the International Stormwater Best Management Practices Database. (www.bmpdatabase.org). It is operated by a clearinghouse that is charged with review of the data being submitted for quality and minimum standards of data integrity. The clearinghouse manages data input and its retrieval, and can assist researchers and practitioners in developing information they may be seeking. The main purpose for this database is to have data available internationally to aid professionals in assessing BMP performance and in the establishment, calibration, or verification of deterministic and statistical models that include BMPs. Another goal is to provide the data that permits investigators to search for quantifiable relationships between BMP designs and their performance regardless of where they may be physically located.

The database's graphical user interface and project data can be downloaded from the project web site or obtained by mail as a CD version. The project web site contains many useful components and project documents, including:

1. International Stormwater Best Management Practices Software – The software's graphical user interface is comprised of two separate components: (1) the data entry module (2) and the data search engine.
2. On-line search engine/data tables – The on-line search engine provides database users with up-to-date BMP study details and statistical summaries

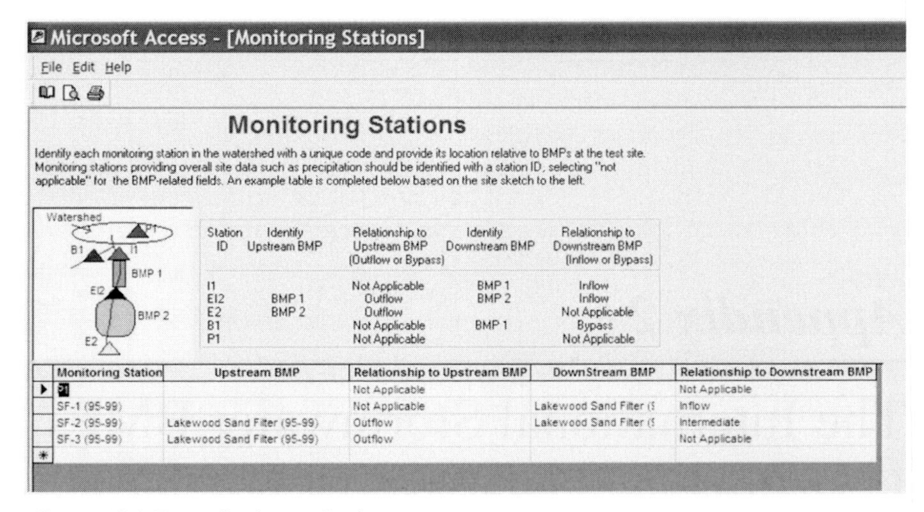

Figure A2.1 Example site on database.

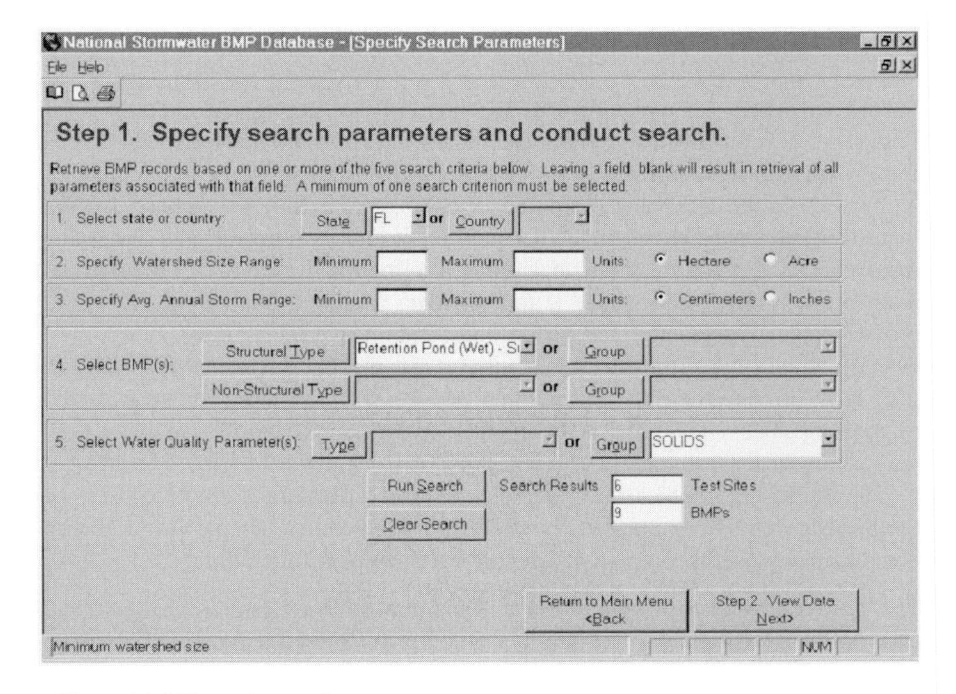

Figure A2.2 Example search page.

for individual test sites. The separate on-line data tables include Excel spreadsheets and text files of (1) individual storm event flow, precipitation and water quality data for structural BMPs in the database and (2) statistical summaries of these storm event data by BMP test site.

3. Separate on-line 'flat-file' data tables include Excel spreadsheets and text files of (1) individual storm event flow, precipitation and water quality data for each structural BMP study in the database and (2) statistical summaries of the storm event data by BMP test site.
4. Project documents – several key project documents are available including:
   a. the software user's manual
   b. the database data elements list,
   c. the initial study review bibliography (779 entries)
   d. the BMP performance monitoring and guidance manual

The information compiled in the database encompasses two basic categories for BMP studies:

1. BMP design, watershed, site information
2. monitoring data (water quality, flow, and precipitation)

Figure A2.3 The Guidance Manual for meeting the National Stormwater BMP Database Requirements.

Table A2.1 Numbers of major types of BMPs in the database.

BMP type	*No of BMPS in category with design information*	*Precipitation records for BMP type*	*Flow records for BMP type*	*Water quality records for BMP type*
Detention basins	24	129	229	4,209
Grass filter strip	32	227	385	6,251
Media filter	30	187	327	6,144
Porous pavement	5	5	5	55
Retention pond	33	378	817	14,293
Percolation trench and dry well	1	3	3	21
Wetland channel and swale	14	53	113	1,241
Wetland basin	15	221	681	7,320
Hydrodynamic devices	16	169	309	6,186
Total	**170**	**1,372**	**2,869**	**45,720**

At the end of 2003 this database contained the structural BMP data shown in Table A2.1.

Currently, the most frequent use of these data is for assessing potential performance of BMPs in terms of the changes in run-off volumes, the amount of run-off treated or not treated, and the effluent quality achieved. The percentage removed can also be calculated from this data, but the database team recommends using effluent event mean concentration or effluent loads as a basis of comparing performance measures between BMPs. In addition, more research-oriented use of the database has included the development, calibration and verification of fundamental process levels within BMPs and 'black-box' statistical models of BMP performance, as well as providing estimates of expected or achievable effluent quality.

Two major ongoing research projects in 2004 are focusing on BMP performance at the unit process level, and are using the database project findings and accumulated monitoring study data as primary sources of BMP effluent event mean concentration levels, environmental and design information.

The first is the National Cooperative Highway Research Program (NCHRP) project 'Evaluation of Best Management Practices for Highway Run-off Control', which is intended to provide evaluation of the basic scientific and technical criteria that can be used for the quantitative assessment of wet-weather flow (WWF) control alternatives for highways and other highway-related facilities, and demonstrate the application of these criteria in facilitating effective implementation of such controls.

The database represents a primary source of data on structural controls in the near-highway environment and is being used in a manner similar to the project approach for the WERF project described below. Where analytical models can be developed, the database is being used as a source of model calibration data marrying the empirical 'black-box' approach of EMC statistical evaluations with intra-event analytical and numerical modelling.

The second of these two projects is the Water Environment Research Foundation (WERF) project 'Critical Assessment of Stormwater Control Selection Issues'. The goal of the WERF project is to provide., as far as possible, guidance for selection and sizing of most stormwater run-off quality controls based on literature review, available data, and fundamental principles of unit operation. As such, the database is a primary source for monitoring data that helps to characterize processes and verify process level models used in the research. Sites contained in the database represent the best available datasets for this purpose. The work includes an evaluation of the ability of some BMPs to reduce the volumes of run-off and therefore reduce pollutant loads as well as reduce physical stream impacts.

Practitioners are often divided on the use of analytical models versus the use of 'black-box' empirical analyses of monitoring data. Both of these projects demonstrate that these approaches are not mutually exclusive and that a combination of analytical models and empirical data analysis may offer the best opportunity to further the understanding of structural stormwater control facilities performance and effects that designs have on performance.

In addition to providing a useful source of detailed monitoring and reporting data for researchers, the database project has generated electronic and hard-copy deliverables useful to stormwater professionals, practitioners, and researchers alike. These reports include individual statistical summaries of water quality, flow, and precipitation data for each BMP in the database. One example of the available statistical summaries is provided in Figure A2.4 and Figure A2.5. These statistical summaries allow selection of an appropriate estimate of the central tendency of effluent quality for a BMP or BMP type (arithmetic estimates of mean effluent concentrations) while accounting for the inherent uncertainty in that estimate. In addition the practitioner is provided with graphical plots as well as standard parametric and non-parametric hypothesis test results and tests of distributional behavior of the data. The information contained on the statistical summaries is also available in a flat file format from the project web site at http://www.bmpdatabase.org/DB_DownLoad/Analysis_Excel.zip. For further information, contact the clearinghouse at http://www.bmpdatabase.org/Contact.htm

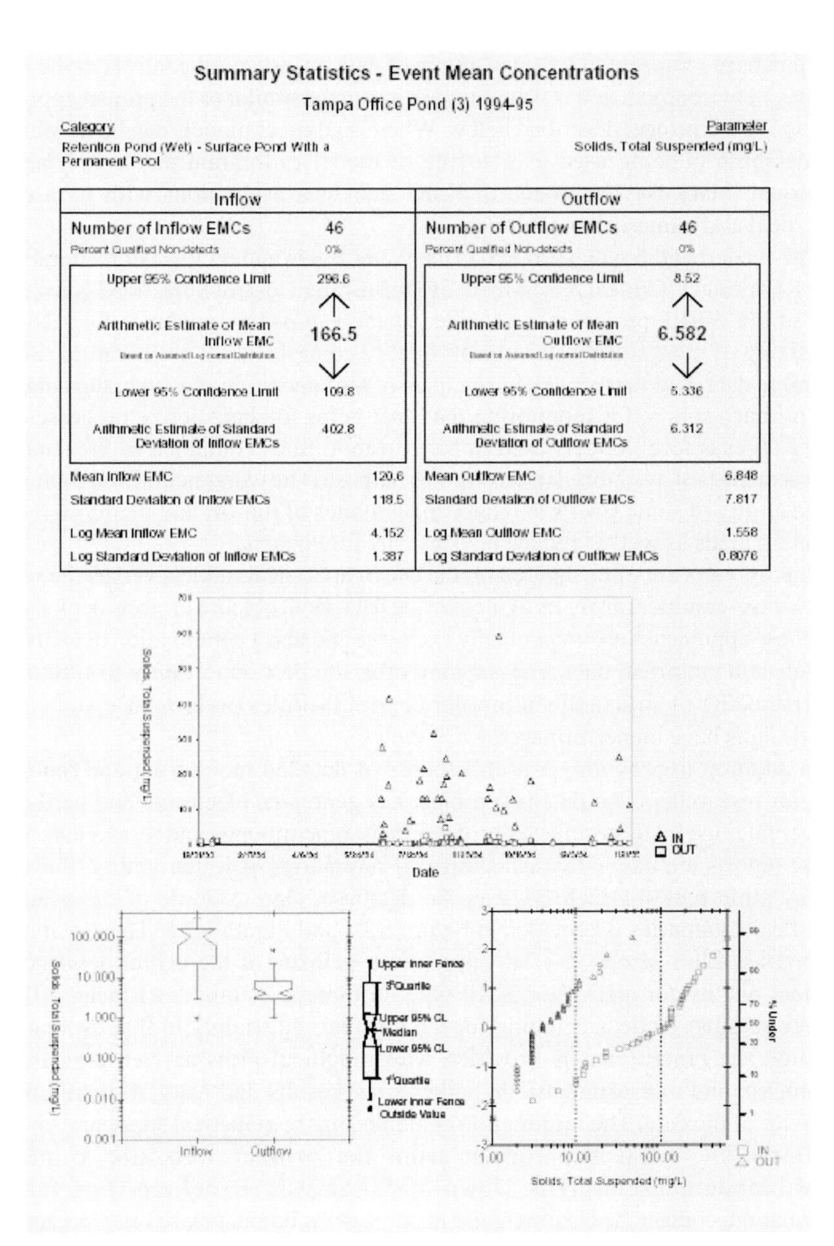

Summary Statistics - Event Mean Concentrations

Tampa Office Pond (3) 1994-95

Category: Retention Pond (Wet) - Surface Pond With a Permanent Pool

Parameter: Solids, Total Suspended (mg/L)

Inflow		Outflow	
Number of Inflow EMCs	46	Number of Outflow EMCs	46
Percent Qualified Non-detects	0%	Percent Qualified Non-detects	0%
Upper 95% Confidence Limit	296.6	Upper 95% Confidence Limit	8.52
Arithmetic Estimate of Mean Inflow EMC (Based on Assumed Log-normal Distribution)	166.5	Arithmetic Estimate of Mean Outflow EMC (Based on Assumed Log-normal Distribution)	6.582
Lower 95% Confidence Limit	109.8	Lower 95% Confidence Limit	5.336
Arithmetic Estimate of Standard Deviation of Inflow EMCs	402.8	Arithmetic Estimate of Standard Deviation of Outflow EMCs	6.312
Mean Inflow EMC	120.6	Mean Outflow EMC	6.848
Standard Deviation of Inflow EMCs	118.5	Standard Deviation of Outflow EMCs	7.817
Log Mean Inflow EMC	4.152	Log Mean Outflow EMC	1.558
Log Standard Deviation of Inflow EMCs	1.387	Log Standard Deviation of Outflow EMCs	0.8076

Figure A2.4: Example of the first page of the statistical summary for total suspended solids for the Tampa Office Pond 1994–1995 from the International Stormwater Best Management Practices Database Project.

Summary Statistics - Event Mean Concentrations

Tampa Office Pond (3) 1994-95

Category	Parameter
Retention Pond (Wet) - Surface Pond With a Permanent Pool	Solids, Total Suspended (mg/L)

Summary of Distributional Characteristics

Shapiro-Wilks W-test (n<50) $\alpha = 0.05$			
Inflow		**Outflow**	
Inflow EMCs Normally Distributed?	No	Outflow EMCs Normally Distributed?	No
Inflow EMCs Log Normally Distributed?	No	Outflow EMCs Log Normally Distributed?	Yes

Lilliefors Test (used when n>50) $\alpha = 0.05$			
Inflow		**Outflow**	
Lillifors Probability for Inflow EMCs	0.006	Lillifors Probability for Outflow EMCs	0
Lillifors Probability for Log Transformed Inflow EMCs	0.004	Lillifors Probability for Log Transformed Outflow EMCs	0.002

Hypothesis Test Results - Raw Data

Nonparametric Analysis - Mann-Whitney Test $\alpha = 0.05$			
Reject the Null Hypothesis that the two means are the same?	Yes	Mann-Whitney Probability	0

Parametric Analysis - t-Test $\alpha = 0.05$			
Separate Probability	0	Pooled Probability	0
Reject the Null Hypothesis that the two means are the same? Assuming Equal Variance.	Yes	Reject the Null Hypothesis that the two means are the same? Assuming Unequal Variance.	Yes

Hypothesis Test Results - Log Transformed Data

Nonparametric Analysis - Mann-Whitney Test $\alpha = 0.05$			
Reject the Null Hypothesis that the two means are the same?	Yes	Mann-Whitney Probability	0

Parametric Analysis - t-Test $\alpha = 0.05$			
Separate Probability	0	Pooled Probability	0
Reject the Null Hypothesis that the two means are the same? Assuming Equal Variance.	Yes	Reject the Null Hypothesis that the two means are the same? Assuming Unequal Variance.	Yes

Test of Equal Variance - Levene Test - Log Transformed Data

Equal Variance?	Yes	Probability	0.001

Disclaimer

This database on which this statistical summary is based is intended to provide a consistent and scientifically defensible set of data on BMP designs and related performance. Although the database team has made an extensive effort to assess the quality of the data entered for consistency and accuracy, the use of the database information or any analyses results provided by the project team is solely at the risk and option of the user. The intended purpose of the database is to provide a data exchange tool that permits characterization of BMPs solely upon their measured performance using the same protocols for measurements and reporting information.

The Database team does not endorse any BMP over another and any assessments of performance by others should not be interpreted or reported as the recommendations of the Database team.

Figure A2.5 Example of the second page of the statistical summary for total suspended solids for the Tampa Office Pond 1994-1995 from the International Stormwater Best Management Practices Database Project

Acknowledgements

Principal Investigators for the Urban Water Resources Research Council (UWRRC):
Ben Urbonas (Denver Urban Drainage and Flood Control District)
Eric Strecker (GeoSyntec Consultants)
Jon Jones (Wright Water Engineers, Inc.)

EPA Project Officers:
Eric Strassler
Jesse Pritts

Other primary project team members:
Marcus Quigley (GeoSyntec Consultants),
Tom Langan, Jane Clary, and John O'Brian (Wright Water Engine Glossary
282282 Diffuse Pollutioners, Inc)

Glossary of diffuse pollution and best management practice

Adjuvants Additives to pesticides to improve their effectiveness without increasing the dose of active chemical.

Attenuation The process of providing a level of storage volume with a restricted outlet within the surface-water drainage system/network to detain the surface water run-off for a period of time particularly during periods of peak flow and therefore control run-off flows.

Attenuation storage volume The level of storage volume with a restricted outlet provided within the surface water drainage system/network to detain the surface water run-off for a period of time particularly during periods of peak flow and therefore control run-off flows.

Balancing pond A pond provided to attenuate surface water run-off flows by storing run-off during peak flow periods and releasing it in a controlled manner after the peak flow has passed.

Bank full flow Is the condition where a burn, stream or river is running at full channel capacity without overcapping of its banks.

Biobed An area underlain by organic matter rich substrates used for filling pesticide sprayers. Designed to safely biodegrade any spilt chemical.

 ISBN: 1 900222 53 1. Published by IWA Publishing, London UK.

BMP (best management practice) A single measure, prcedure or structure to control an aspect of diffuse pollution (plural: BMPs)

Buffer strip Strip of grass or trees between farmed or urban land and water course.

Catchment The area set by the topography of the surrounding land, which naturally contributes surface water run-off flow to a point on a watercourse or drainage system and is used to calculate run-off flows and volumes.

Conservation tillage One of a number of methods of cultivating land aimed at reducing soil loss by erosion. Most methods involve leaving the residues of the previous crop on the surface to protect the soil.

Crop rotation Farming method which varies the crop grown in any one field each year. May help to reduce pesticide use and reduce average rates of soil erosion.

Curtilage The general land area within the development property or properties boundaries.

Design criteria The design standards set by the regulators and planners, normally in discussion with the developer, that the proposed surface water drainage system must satisfy.

Detention basin A dry basin designed to temporarily store and attenuate surface water peak flows.

Diffuse pollution 'Pollution arising from land-use activities (urban and rural) that are dispersed across a catchment or subcatchment, and do not arise as a process industrial effluent, municipal sewage effluent, deep mine or farm effluent discharge.'

Diversion terraces An erosion-control system comprising a series of shallow broad-based ditches across fields designed to reduce the volume of surface stormwater run-off by breaking the slope into shorter lengths.

Drainage system/network The network of swales, channels, drainpipes, filter drains and trenches, incorporating SUDS control and treatment devices to collect, manage and convey surface water run-off through and out of the development site.

Extended detention basin A detention basin where the outflow control is restricted beyond that allowed purely for attenuation so that the surface water run-off is detained longer allowing for some increased removal of particulate pollutants by settlement.

Filter drain Basically a linear trench filled with granular filter material, usually with a perforated drain, in the bottom of the trench.

First flush run-off As the name implies, this is the first part of the surface water run-off to flow over and effectively wash the development surfaces in the early stages of a storm event and it therefore tends to have more concentration of pollutants than the remainder of the surface water run-off.

Flood risk assessment A formalized report investigating the potential for flooding both within the development and upstream and downstream of it.

Floodplain An area of low-lying land adjacent to a burn, stream or river, which is liable to regular flooding when the flow in the burn, stream or river exceeds its capacity.

Flow control devices These are generally mechanical devices such as restricted orifice plates and pipes, slotted and V-notch weirs, perforated risers and proprietary vortex-type devices, e.g. hydro-brakes, which are normally sited either at the outlet from the development site or as part of the outlet from a SUDS facility/structure.

Flow velocity Basically the average speed at which surface water run-off flows in a channelized or piped drainage network.

Grassed waterway A grassed depression down a field designed to safely carry away stormwater run-off without causing erosion.

Green manure A green crop grown, not to be harvested, but to be ploughed in to protect soil and increase soil organic mater content.

Growth gurves These are used in hydraulic calculations to assess the appropriate multipliers required to convert a basic result into a range of results relating to the various storm return periods.

Hydraulic gradient Generally the ratio of the head loss to distance and can be directly related to the flow velocity.

Impermeable and **permeable** These are the terms used to describe whether surfaces, materials and ground strata are respectively incapable or capable of allowing surface water run-off to drain into and through them and are therefore candidates for infiltration or not.

Infiltration devices Soakaways, infiltration trenches and basins are devices which encourage the surface water to infiltrate into the ground when the nature of the soils and substrata permit, i.e. sands and gravels.

Infiltration to the ground The process of passing surface water through the surface and into the ground strata beneath.

Infiltration in a sewer or drain The entry/seepage of groundwater into the sewer or drain.

Integrated pest management One of various methods of reducing pesticide usage by encouraging natural predators and accurately targeting pesticide inputs at the specific pest.

Interflow Shallow infiltration of the surface water run-off into the upper soil strata where it can flow slowly through the soil layers or be held and subsequently evaporate.

M5-60 A rainfall event with a return period of 5 years, or expectancy of the event occurring in any year being 1 : 5 or 20%, and with a duration of 60 minutes.

Minimum tillage Methods of establishing crops without using a plough or other deep tillage implement. Often part of an erosion control strategy.

Nitrification inhibitor One of various chemical additives to fertilizers and manures designed to prevent or slow the natural transformation from ammoniacal nitrogen to nitrate nitrogen. May reduce nitrate leaching.

NPS (non-point source pollution) A legal term in the USA encompassing minor point sources, plus strictly non-point in a geographic sense too. Superseded, therefore, by **diffuse pollution**.

Nutrient balance sheet A calculation of the relative quantities of any given nutrient coming into an area in the form of fertilizers, manures, feeds etc compared with the quantities leaving in the form of harvested crop, sold milk, stock etc. May be done on a field, whole-farm or catchment scale.

Open channel capacity Is the assessed channel flow for the channel being full of run-off water, i.e. bank full.

Open channel flow The assessed flow for a burn, channel, stream or river normally taken as an idealized trapezoidal channel using Manning's formula.

Permeable and **impermeable** These are the terms used to describe whether surfaces, materials and ground strata are respectively capable or incapable of allowing surface water run-off to drain into and through them and are therefore candidates for infiltration or not.

Plate/pipe orifice control Flow control device normally set into a weir wall within a manhole/chamber to provide restricted outflow. If this is combined with a storage facility that has a reasonable head of water range then it can provide some variable outflow control.

Point source. In pollution control terms, a continuous discharge of effluent to a water body. Most usefully used to refer to major point sources, such as industrial process effluents, and municipal sewage discharges.

Porosity A characteristic of natural soil strata that can be tested and numerically assessed by a set standard test to indicate the infiltration potential of the ground as tested.

Porous and permeable paving Areas surfaced with materials that allow the surface water to pass through into the sub-base storage void beneath.

Precision farming Farming methods where soil nutrient levels and crop yields are mapped across fields and fertilizer rates adjusted accordingly. Usually uses GPS technology. Results need careful interpretation to be useful.

Rainfall intensity Indicates the average level of rainfall in mm/hr.

Regional SUDS Generally deal with major developments or multiple developments with combined areas or catchments between 5 ha and upwards of 50 ha depending on the natural topography and contours.

Resident pond time The effective equivalent time which surface water run-off entering a retention pond or wetland could be said to take to flow through the facility. This prolonged time allows for biological treatment.

Retention pond A permanently wet pond with perimeter aquatic planted vegetation where the surface water run-off enters and is retained for a sufficient number of days to allow settlement and biological treatment to occur.

Riparian wetland A linear wetland adjacent to a watercourse. Believed in some circumstances to improve removal of soluble pollutants such as nitrates.

Run-off coefficient A term normally expressed as a percentage used in surface water run-off calculations to represent the impermeable portion of the development or catchment.

SAAR Standard annual average rainfall in millimetres.

Scheduling irrigation Methods of matching rates of application of irrigation water to crop water usage. May both conserve water and reduce pollution risks.

Soil Winter rain acceptance potential (WRAP) soil classification coefficient.

Site control SUDS Generally deal with medium development site catchment and sub-catchment areas.

Source control SUDS As the name implies, controls/treats surface water run-off at or near its source.

Storm duration The time from the start to the finish of the storm event rainfall.

Storm rainfall hydrograph Graphical representation of a storm event with run-off flow being plotted against time.

Storm return period The statistically average period in years within which a specific storm event is likely to return.

Storm run-off peak flows/discharges The maximum or peak run-off flows generated by a storm event. Run-off flows are normally quoted in either cumecs (cubic metres per second) or l/s (litres per second).

Storm run-off volume The total assessed volume of surface water run-off generated during a storm event.

Stormwater run-off Water normally generated by rainfall during storm events, which flows over impermeable or saturated permeable ground surfaces to discharge to a watercourse or collector drainage system.

Sub-catchment A division or part of the full catchment, which is considered separately in terms of surface water run-off control and/or treatment.

SUDS (sustainable urban drainage system) A single or sequence of surface water drainage management practices providing quantity control and/or quality treatment to the surface water run-off in a more sustainable manner than conventional drainage methods.

Surface water drainage management The management of the surface water run-off as it flows through and out of the development site.

Surface water run-off Water normally generated by rainfall during storm events, which flows over impermeable or saturated permeable ground surfaces to discharge to a watercourse or collector drainage system.

Swale A shallow grass lined channel used to treat, convey and, on occasion, control surface water run-off.

Time of concentration The time notionally taken for rain/stormwater falling at the extremities of a catchment to run across the surface and enter the point of the drainage network or watercourse under consideration.

Treatment volume The volume of surface water run-off, which is regarded as containing the most polluted part of the surface water run-off from any storm. It is normally set at a 90 per cent capture level of all annual storms, which effectively means that it will cater for all small storms and the first flush run-off from the larger storms.

V_t Treatment volume (see preceding entry)

Watercourse Any natural burn, stream, river or artificial channel that conveys surface water.

Wetland A permanently wet pond, similar to a retention pond in design and function, but very much shallower with far greater aquatic planted vegetation cover and far less open water.

Zero tillage Methods of sowing new crop directly into the residues of the previous crop without any cultivation.

References

Abbott CL and Comino-Mateos L (2003) *In-situ* hydraulic performance of a permeable pavement sustainable urban drainage system. *Water and Environmental Management Journal*, **17**(3): 187–90.

Abujamin S and Abujamin S (1985) Crop residues mulch for conserving soil in uplands of Indonesia. In SA El-Swaify, WC Moldenhauer, and A Lo, eds *Soil Erosion and Conservation*, 607–14, Soil Conservation Society of America, Ankeny, IA.

Adiscott TM (1997) A critical review of the value of buffer zone environments as a pollution control tool. In Haycock *et al* (eds) *Buffer Zones: Their Processes and Potential in Water Protection.* Quest Environmental.

Agrawal, G.D (1999) Diffuse agricultural pollution in India. (Paper presented at the International Workshop, Teolo, Italy). *Water Science and Technology,* **39**(3):14.

Agrawal GD, Lunkad SK and Malkhed T (1999) Diffuse agricultural nitrate pollution of groundwaters in India. *Water Science and Technology,* **39**(3): 67–75.

Aitken M, Merrileees D, Lewis D and Jones A (2003) Impact of agricultural practices and catchment characteristics on Ayrshire bathing waters. In I McTaggert and L Gairns (eds) *Agriculture, Waste and the Environment* Scottish Agricultural College (SAC) Edinburgh.

Alcock RE, Boumphrey R, Malcolm HM, Osborn D and Jones KC (2002) Temporal and spatial trends of PCB congeners in UK gannet eggs. *Ambio* **31**(3).

Anderson C (1999) Integration of broad-based community values in urban stormwater management. In AC Rowney, P Stahre and LA Roesner (eds) *Sustaining Urban Water Resources in the 21st Century*. Proceedings of an Engineering Foundation conference,

September 7-12 1997, Malmö, Sweden. The American Society of Civil Engineers, Reston, VA.

Andoh R, Stephenson A and Kane A (2000) Sustainable urban drainage using the Hydro Stormcell storage system. In CJ Pratt (ed.) *Proceedings Standing Conference on Stormwater Source Control,* Vol. XIX. School of The Built Environment, Coventry University, Coventry.

Apostolaki S, Jefferies C and Souter N (2001) Assessing the public perception of SUDS at two locations in eastern Scotland. In CJ Pratt, JW DAvies and JL Perry (eds) *Proceedings of the 1st National Conference on Sustainable Drainage.* 18–19 June, Coventry University.

Arheimer B, Andersson L, Hansson LA, Joborn A, Lindstrom G, Olsson J and Pers BC (2002) Modelling diffuse nutrient flow in eutrophication control scenarios. In *Proceedings of the 6th International Conference on Diffuse Pollution*, Amsterdam, 30 September–4 October. International Water Association, in co-operation with the Netherlands Association on Water Management (NVA) and Aquatech, Amsterdam.

ASCE-WEF Joint Task Force (1998) *Urban Runoff Quality Management.* WEF Manual of Practice No. 23, Water Environment Federation, Alexandria, VA.

Avcievala, S (1991) *The Nature of Water Pollution in Developing Countries.* Natural Resources Series No. 26, UNDTCD, United Nations, New York.

Baardwijk FAN van and Spijker JH (2002) Policies for conversion to non-chemical control at municipalities. In *Proceedings of the 6th International Conference on Diffuse Pollution*, Amsterdam, 30 September–4 October. International Water Association, in co-operation with the Netherlands Association on Water Management (NVA), Amsterdam.

Baldwin DJ, Davies K, Sanders G.B, and Nicholson RJ (2003) Implementing a waste minimisation strategy in agriculture. In *Agriculture, Waste and the Environment: Selected Papers from the SAC/SEPA Biennial Conference.* Edinburgh, 26–28 March 2002. SAC Publishing, Edinbrugh.

Balmforth D, Percy C and Bray R (2003) A risk management approach to adoption of sustainable drainage. *Proc. Second National Conference on Sustainable Drainage.* Coventry University, Coventry, 23–24.

Bannink, AD (2002) How Dutch drinking water is affected by the use of herbicides on pavements. In *Proceedings of the 6th International Conference on Diffuse Pollution*, Amsterdam, 30 September–4 October. International Water Association, in co-operation with the Netherlands Association on Water Management (NVA) and Aquatech, Amsterdam.

Bateman M, Livingston EH and Cox J (1998) *Overview of Urban Retrofit Opportunities in Florida.* Florida Department of Environmental Protection, Tallahasee, FL.

Bauer B, Fiorini P, Schulte-Oehlmann U, Oehlmann J, and Kalbfus W (1997) The use of *Littorina littorea* for tributyltin (TBT) monitoring – results from German TBT survey 1994/1995 and laboratory experiments. *Environmental Pollution*, **96**(3):299–309.

Beck KC, Reuter JH and Perdue EM (1974) Organic and inorganic geochemistry of some coastal plain rivers of the southeastern United States. *Geochimica et Cosmochimica Acta* **38:**341-364.

Bell W, and Chanpagne PC (1998) Targets of opportunity: Alexandria's urban retrofit program. National Conference on Retrofit Opportunities for Water Resource Protection in Urban Environments, February, Chicago, IL.

Bendoricchio G, Calligaro L and Carrer G (1999) Consequences of diffuse pollution on the water quality of rivers in the watershed of the Lagoon of Venice (Italy). *Water Science and Technology* **39**(3):113–20.

Bendoricchio G, Burigana E, Calligaro L and Carrer G (2002) Water Quality in an agricultural watershed. In *Proceedings of the 6th International Conference on Diffuse Pollution*, Amsterdam, 30 September–4 October. International Water Association, in co-operation with the Netherlands Association on Water Management (NVA) and Aquatech, Amsterdam.

Birdlife International (2002) *CAP Reform Proposals.* The RSPB, Sandy, UK.

Blevins RL, Lal R, Doran JW, Langdale GW and Frye WW (1998) Conservation tillage for erosion control and soil quality. In FJ Pierce and WW Frye (eds) *Advances in Soil and Water Conservation*, Sleeping Bear Press Inc, Michigan.

Boyd J (1998) Searching for the profit in pollution prevention: case studies in the corporate evaluation of environmental opportunities. Discussion Paper 98-30, Resources for the Future, Washington, DC.

Braskerod BC (2003) Sedimentation in small constructed wetlands. Doctoral thesis, Department of Soil and Water Sciences, Agricultural University of Norway.

Braune B, Muir D, DeMarch M, Gamberg M, Poole K, Currie R and Dodd M (1999) Spatial and temporal trends of contaminants in Canadian Arctic freshwater and terrestrial ecosystems: a review. *Science of the Total Environment*, **230**(1–3): 145–208.

Bray R (2000) Design of the treatment rain at Hopwood Park, MSA M42. *Proceedings of the Standing Conference on Stormwater Source Control.* School of The Built Environment, Coventry University, Coventry.

Bray R (2001a) Environmental monitoring of sustainable drainage at Hopwood Park Motorway Service Area M42 Junction 2. In C Pratt (ed.) *Proceedings of First National Conference on Sustainable Drainage.* School of The Built Environment, Coventry University, Coventry.

Bray R (2001b) Maintenance of sustainable urban drainage: experience on two EA demonstration sites in England. In C Pratt (ed.) *Proceedings of First National Conference on Sustainable Drainage.* School of The Built Environment, Coventry University, Coventry.

Bray R (2003) Sustainable drainage solutions for local authority school sites. *Proc. Second National Conference on Sustainable Drainage*, Coventry University, 23–24 June, CIRIA, UK.

Bright DA and Ellis DV (1989) A comparative survey of Imposex in north-east Pacific neogastropods (*Prosobranchia*) related to tributyl tin contamination, and the choice of suitable bioindicator. *Canadian Journal of Zoology*, **68**: 1915–24.

Brown L (1976) *British Birds of Prey.* Collins New Naturalist Series, 1989 edition. Bloomsbury Books, London.

Brundtland Commission (1987) *Our Common Future.* The World Commission on Environment and Development.

Burkart M and Heath R (eds) (2002) Diffuse pollution and watershed management. *Water Science and Technology* **45**(9).

Cambell, NS (1999) Best management practice in urban stormwater drainage: practical implementation of the BMP approach. *Water Course: Urban Catchment Drainage.* IEI (Irish Engineers Institute), Dublin.

Camp Dresser and McKee, Inc. (1993) *California Best Management Manual*, Camp Dresser & McKee Inc., Orlando, FL.

Campbell, CS and Ogden MH (1999) *Constructed Wetlands in the Sustainable Landscape*. John Wiley & Sons Inc., New York.

Carlsson H (2001) personal communication, based on presentation at EU workshop in Brussels, 12 October 2001.

Carson R (1962) *Silent Spring*. Houghton-Mifflin, New York.

Carter MR 1998. Conservation tillage practices and diffuse pollution. In *Diffuse pollution and agriculture II*. The Scottish Agricultural College, Aberdeen.

CEC (1994) *Environmental Programme for the Danube River Basin, Danube Integrated Environmental Study*. Phase I. Commission for European Communities.

CES (1993) *Uses, Fates and Entry into the Environment of Nonylphenol Ethoxylates*. Consultants in Environmental Sciences Ltd., London.

CH2M-Hill (1990) *Non-point Source Impact Assessment*, Water Environment Research Foundation, Alexandria, VA.

Chambers JM, Wrigley TJ and McComb AJ (1993) The potential use of wetlands to reduce phosphorus export from agricultural catchments. *Fertilizer Research* **36**: 157–64.

Chambers BJ, Smith KA and Pain BF (2000) Strategies to encourage better use of nitrogen in animal manures. *Soil Use and Management* **16:** 157–61.

Chave, P. (2001) *The EU Water Framework Directive*. IWA Publishers, London.

Choe JS, Bang KW and Lee JH (2002) Characterisation of surface runoff in urban areas. *Water Science and Technology* **45**(9): 249–54.

Christian DG, Goodlass G and Powlson DS (1992) Nitrogen uptake by cover crops. *Aspects of Applied Biology* **30**: 291–300.

CIRIA (2000) *Sustainable Urban Drainage Systems: Design Manual for Scotland & Northern Ireland*. Construction Industry Research and Information Association Report no. C521. CIRIA, London.

Clark, EH, Havencamp, JA and Chapman W (1985) *Eroding Soils: The Off-farm Impact*. The Conservation Foundation, Washington, DC.

Clary J, Urbonas B, Jones J, Strecker E, Quigley M and O'Brien J (2001) Developing and evaluating a stormwater BMP database. In Novatech 2001 *Innovative Technologies In Urban Drainage*, 4th International Conference, 25–27 June, Lyon, France. *Water Science and Technology* **45**(7).

Cole RH, Frederick RE, Healy RP, and Rolan, RG (1984) Preliminary findings of the priority pollutant monitoring project of the nationwide urban runoff program. *Journal of the Water Pollution Control Federation*, **56**(7): 898–908.

Committee to Assess the Scientific Basis of the TMDL Program (2001) *Assessing the TMDL Approach to Water Quality Management,* National Academy Press, Washington, DC.

Committee on Long Range Soil and Water Conservation (1993) *Soil and Water Quality: An Agenda for Agriculture,* National Academy Press, Washington, DC.

Conlin J (2000) Developments in sustainable urban drainage in Scotland. In *Proceedings Standing Conference On Stormwater Source Control Quantity and Quality*. Vol. XIX,. The School of The Built Environment, Coventry University, Coventry.

Conservation Technology Information Centre (CTIC) 1996. *National Crop Residue Management Survey*. CTIC, West Lafayette, IN.

Cunningham, PA (1988) *Non-Point Source Impacts on Aquatic Life: Literature Review.* Prepared for Monitoring and Data Support Division, Office of Water Regulations and Standards, U.S. Environmental Protection Agency. Research Triangle Institute, Research Triangle Park, NC.

D'Arcy BJ (1991) Legislation and the control of dye house pollution. *JSDC,* **107**: 387–9.

D'Arcy BJ (1998a) Report on Technical Visit to Canada (Alberta and British Columbia) June 1998. Unpublished technical report No. EQI/IV/6.98 (revised Jan 99). Scottish Environment Protection Agency, Edinburgh.

D'Arcy BJ (1998b) A new Scottish approach to urban drainage in the developments at Dunfermline. *Proceedings of the Standing Conference on Stormwater Source Control.* Vol. XV. The School of the Built Environment, Coventry University, Coventry.

D'Arcy RW (2002) personal communication – unpublished surveys of SUDS ponds for water voles.

D'Arcy BJ and Bayes CD (1994) Industrial estates a problem. In Pratt CJ (ed.) *Stormwater Management on Industrial Developments. Proceedings of the Standing Conference on Stormwater Source Control,* Vol. X, The School of the Built Environment, Coventry University, Coventry.

D'Arcy BJ and Frost CA (2001) The role of best management practices in alleviating water quality problems associated with diffuse pollution. *Science of the Total Environment* **265**: 359–67.

D'Arcy BJ and Harley D (2002) Driving regulatory reforms for sustainable catchment management. International Water Association Leading Edge Conference 'Sustainability in the Water sector', 25–26 November, Venice, Italy. (Abstracts and CD of presentations from IWA, London.)

D'Arcy BJ and Roesner LA (1999) Scottish experiences with stormwater management in new developments. In AC Rowney, P Stahre and LA Roesner (eds) *Sustaining Urban Water Resources in the 21st Century.* Proceedings of an Engineering Foundation conference, 7–12 September 1997, Malmö, Sweden. The American Society of Civil Engineers, Reston, VA.

D'Arcy BJ, Usman F, Griffiths D and Chatfield P (1998) Initiatives to control diffuse pollution in the UK. *Water Science and Technology* **38**(10): 131–8.

D'Arcy BJ, Ellis JB, Ferrier RC, Jenkins A and Dils R (2000a) *Diffuse Pollution Impacts: The Environmental and Economic Impacts of Diffuse Pollution in the UK.* Chartered Institute of Water and Environmental Management (CIWEM), Terence Dalton Publishers, Lavenham.

D'Arcy BJ, MacCalman K, Marsden M, and Ellis JB (2000b) The control of diffuse pollution from industrial areas. Paper presented at the IAWQ 4th International conference on Diffuse Pollution, 16–21 January, Bangkok. International Water Association, London.

Davies DHK and Christal A (1996) Buffer strips: a review. In *Diffuse Pollution and Agriculture,* 183–92, The Scottish Agricultural College, Aberdeen.

De Wit J (1994) *Civic Environmentalism – Alternatives to Regulation in States and Communites.* Congressional Quarterly, Inc.

DEFRA (2002) *The Environment In Your Pocket: Key Facts And Figures On The Environment Of The United Kingdom.* Department for Environment, Food and Rural Affairs (DEFRA Publications) , London.

Deletic A (1998) The first flush load of urban surface runoff. *Wat. Res.*, **32**(8): 2462–70.

Dennison W (2002) Effective responses to coastal water management challenges: lessons from Australia and the USA. International Water Association Leading Edge Conference 'Sustainability in the Water sector' 25–26 November, Venice, Italy. (Abstracts and CD of presentations from IWA, London.)

Dennison WC and Abal EG (1999) *Moreton Bay Study: A Scientific Basis For The Healthy Waterways Campaign.* South East Queensland Regional Water Quality Management Strategy, Brisbane City Coucil, Brisbane.

Department of the Environment (1983) *Design and Analysis of Urban Storm Drainage – The Wallingford Procedure.* Vols 1–7. National Water Council, Standing Technical Council on Sewers and Water Mains. Report No. 28. Hydraulic Research Ltd, Wallingford, UK.

Dillaha TA, Sherrard JH, Lee D, Mostaghimi S and Shanholtz VO (1988) Evaluation of vegetative filter strips as a best management practice for feed lots. *Journal of the Water Pollution Control Federation* **60**: 1231–8.

Dillaha TA, Reneau RB, Mostaghimi S, and Lee D (1989a) Long-term effectiveness of vegetative filter strips. *Water, Environment and Technology*, **1**: 419–21.

Dillaha TA, Reneau RB, Mostaghimi S and Lee D (1989b) Vegetative filter strips for agricultural non-point source pollution control. *Transactions of the American Society of Civil Engineers*, **32**: 513–19.

Division of Drinking Water Quality Control (1993) *Implications of Phosphorus Loading for Water Quality in NYC Reservoirs.* NYC Department of Environmental Protection, New York

Dorman ME, Hartigan J, Johnson F and Maestri B (1988) *Retention, Detention and Overland Flow for Pollutant Removal from Urban Stormwater Runoff* FHWA/RD–87/056. Federal Highway Administration, McLean, VA.

Duff-Brown, B. (1999) A 'lost tribe' fights a losing battle. *Milwaukee Journal-Sentinel*, 15 August: 17A, 36A.

DWAF (1998) *Proposal for a national strategy for managing the water quality effects from dense settlements: proceedings of a national consultative workshop to discuss the proposed strategy*. Department of Water Affairs and Forestry, Pretoria.

DWAF (2002) *Water Services Information Systems (Intranet).* Department of Water Affairs and Forestry, Pretoria.

Dwyer J, Baldock D and Einschutz S (2000) *Cross-compliance under the Common Agricultural Policy*. Institute for European Environmental Policy, London.

ECE (1992) *Protection of Inland Water against Eutrophication.* Paper ECE/ENVWA/26. United Nations Economic Commission for Europe, Geneva.

Edwards AC and Withers PJA 1998. Soil phosphorus management and water quality: a UK perspective. *Soil Use and Management* **14**: 124–30.

EEA (1999) *European Environment at the Turn of the Century.* European Environment Agency, Copenhagen.

Ellis JBE (1982) Blights and benefits of urban stormwater quality control. In RE Featherstone and A James (eds) *Urban Drainage Systems*. Pitmans Books Ltd, London.

Ellis JB (1989) *Urban Discharges and Receiving Water Quality Impacts*. Pergamon Press, Oxford.

Ellis JB (1996) Oestrogenic substances in sewage treatment works effluent. In F Seiker and HR Verworn (eds) *Urban Storm drainage: Proceedings of the 7th International IAWQ Conference*. SuG-Verlagsgesellschaft, Hanover.

Ellis, JB and Chatfield P (2001) Diffuse urban oil pollution in the UK. Presented at 5th International Conference on Diffuse Pollution and Watershed Management, International Water Association, Milwaukee, and in *Urban Water* 15–20 June 2001.

Ellis JB, D'Arcy BJ and Chatfield PR (2002) Sustainable urban drainage systems and catchment planning. *JCIWEM*, **16**: 286–91.

England G (1998) Maintenance of stormwater retrofit projects. Available online at http://www.stormwater-resources.com/library.htm#bmps

England G, Dee D and Stein S (2000) Stormwater retrofitting techniques for existing development. Available online at http://www.stormwater-resources.com/library.htm#bmps

Environment Agency (1997) *EA report*. 1 May 1997. The Environment Agency, Bristol.

Environment Agency (1999) *Understanding Rural Land Use*. The Environment Agency, Bristol.

Environment Agency (2002a) *Agriculture And Natural Resources: Benefits, Costs And Potential Solutions*. Environment Agency, Bristol.

Environment Agency (2002b) *Interim Guidance on the Disposal of Contaminated Soils* 2nd edn. Enivronmental Agency, Bristol, UK.

Environmental Law Institute (1998) *Almanac of Enforceable State Laws to Control Non-point Source Water Pollution*. ELI, Washington, DC.

Fairhust WA and Partners (2002) *Sustainable Urban Drainage Systems (SUDS) and Watercourse Enhancement in the South East Wedge Development Final Report to Project Group, February 2002*. SIStech (Scottish Institute of Sustainable Technology), Heriot-Watt University, Edinburgh.

Farmer A, Shaw K, Petersen J-E, Baldock D and Newcombe J (2000) *The EU Framework Directive: Controlling Diffuse Agricultural Pollution in Scotland*. Institute for European Environmental Policy, London.

Feldwisch and Frede (1995) Maßnahmen zum verstärkten Gewasserschutz im Verursacherbereich Landwirtschaft. Bonn, *DVWK- Materialien* **2** 1995.

Ferguson BK (1998) *Introduction to Stormwater: Concept, Purpose, Design*. John Wiley & Sons, Chichester.

Fleischer S, Gustafson A, Joelsson A, Pansar J and Stibe L (1994) Nitrogen removal in created ponds. *Ambio* **23**(6): 349–57.

Flint D (1998) The Loch Leven catchment management project: an holistic approach to diffuse pollution. In T Petchey, B D'Arcy and A Frost (eds) *Diffuse Pollution and Agriculture II* The Scottish Agricultural College, Aberdeen.

Flint D (1999) *The Loch Leven Catchment Management Plan*. Scottish Natural Heritage, Perth.

FAO (1990) *Strategy on Water for Sustainable Agricultural Development*. The United Nations, Food and Agriculture Organisation, Rome.

Forestry Authority (2002) *The Forest and Water Guidelines*, 3rd edn. Forestry Authority, London.

Frederick RE and Dressing SA (1993) Technical guidance for implementing BMPs in the coastal zone. *Water Science and Technology*, **28**(3–5): 129–35.
Frost CA (1996) Loch Leven and diffuse pollution. In AM Petchey, BJ D'Arcy and CA Frost (eds) *Diffuse Pollution and Agriculture* The Scottish Agricultural College, Aberdeen.
Frost CA (1999) A summary of the agricultural issues raised at the conference (IAWQ 3rd International Conference on Diffuse Pollution). *Water Science and Technology* **39**(12): 361–64.
Frost CA, Speirs RB and McLean J (1990) Erosion control for the UK: strategies and short-term costs and benefits. In J Boardman, IDL Foster and JA Dearing (eds) *Soil Erosion on Agricultural Land*. John Wiley and Sons Ltd, Chichester.
Frost CA, Stewart S, Kerr D, MacDonald J and D'Arcy BJ (2002) Agricultural environmental management: case studies from theory to practice. In *Proceedings of the 6th International Conference on Diffuse Pollution*, Amsterdam, 30 September–4 October. International Water Association, in co-operation with the Netherlands Association on Water Management (NVA) and Aquatech, Amsterdam.
Fujii S, Tanaka H and Somlya I (2002) Quantitative comparison of forests and other areas with dry weather on input loading in Lake Biwa catchment area. *Water Science and Technology* **45**(9):183–93.
Fujita S (1994a) Infiltration structures in Tokyo. *Water Science and Technology* **30**(1): 33–41.
Fujita S (1994b) Japanese experimental sewer system and the many source control developments in Tokyo and other cities. In CJ Pratt (ed.) *International Perspectives on Stormwater Management. Proceedings of Standing Conference on Stormwater Source Control.* Vol. VIII. School of The Built Environment, Coventry University, Coventry.
Fujita S (1997) Measures to promote stormwater infiltration. *Water Science and Technology* **36**(8–9): 289–93.
Garden M (2003) personal communication.
Gouman E (2002) Reduction of zinc emissions from buildings: the policy of Amsterdam. In *Proceedings of the 6th International Conference on Diffuse Pollution*, Amsterdam, 30 September–4 October. International Water Association, in co-operation with the Netherlands Association on Water Management (NVA) and Aquatech, Amsterdam.
Greiner RW and de Jong J (1984) *The Use of Marsh Plants for the Treatment of Wastewater in Areas Designated for Recreation and Tourism*. RIJP report no. 225. Lelystad.
Groves SJ and Bailey RJ (1997) The influence of sub-optimal irrigation and drought on the crop yield, N uptake and risk of N leaching from sugar beet. *Soil Use and Management* **13**: 190–5.
Gupta, MK (1981) *Constituents of Highway Runoff*, 6 Volumes, FHWA/RD-81/042-047, Federal Highway Administration, Washington, DC.
Gupta PK (1983) Toxicology of pesticides: a review of the problem. Indian Academy of Environmental Biologists 'Effects of pesticides on Aquatic Fauna' conference, Mhow, 18–22 June.
Haiping Z and Yamada K (1996) Estimation for urban runoff quality modelling. *Water Science and Techology* **34**(3–4): 49–54.
Harremoes P (2002) Challenges to implementing sustainability: values, intentions, uncertainty and ignorance. Keynote presentation at International Water Association

Leading Edge Conference 'Sustainability in the Water sector', 25–26 November, Venice, Italy. (Abstracts and CD of presentations from IWA, London.)
Harrington R, Dunne E, Carroll P, Keohane J and Ryder C (2004) The Anne Valley Project: the use of integrated constructed wetlands (ICWs) in farmyard and rural domestic waste waters. In D Lewis and L Gairns (eds) *Water Framework Directive and Agriculture: Proceedings of the SAC and SEPA Biennial Conference*, 24–25 March, Edinburgh. SAC, Edinburgh.
Harris GL, Bailey SW and Mason DJ (1991) The determination of pesticide losses to water courses in an agricultural clay catchment with variable drainage and land management. In *Proceedings Brighton Crop Protection Conference on Weeds.* British Crop Protection Council, Farnham.
Hart K (2003) Personal communication, sales of Formpave Aquablocks in the UK. Formpave, Coleford, Glos., UK.
Hartigan JP (1989) Bassis of design of wet detention BMPs. In LA Roesner, B Urbonas and MB Sonnen (eds) *Design of Urban Runnoff Quality Controls*. ASCE, New York.
Haycock NE, Burt TP, Goulding KWT and Pinay G (eds) (1997) *Buffer Zones: Their Processes and Potential in Water Protection.* Quest Environmental, Harpenden, UK.
Haygarth PM, Chapman PJ, Jarvis SC and Smith RV (1998) Phosphorus budgets for two contrasting grassland farming systems in the UK. *Soil Use and Management* **14**: 160–7.
Heal KV and Drain SW (2003) Sedimentation and sediment quality in sustainable urban drainage systems. *Proceedings Second National Conference on Sustainable Drainage,* Coventry University, 23–24 June.
Heath RGM (2000) The impact of an African megacity on the water resources and local economy. In GD Agrawal and N Tonmanee (eds) *Proceedings of a Scientific Workshop on Diffuse Pollution in Subtropical and Tropical Developing Countries and its Abatement*, Bangkok, Thailand, 16–21 January. International Water Association, UNEP (Nairobi) and Land Development Department, Thailand.
Heath RGM, Hinsch M and Pulles W (2002) Unique implications of policies and solutions to diffuse pollution management in a developing country – South Africa. In *Proceedings of the 6th International Conference on Diffuse Pollution*, Amsterdam, 30 September–4 October. International Water Association, in co-operation with the Netherlands Association on Water Management (NVA) and Aquatech, Amsterdam.
Heathwaite AL, Griffiths P and Parkinson RJ (1998) Nitrogen and phosphorus in runoff from grassland with buffer strips following application of fertilizers and manures. *Soil Use and Management* **14**: 142–8.
Helander B, Olsson A, Bignert A, Asplund L and Litzen K (2002) The role of DDE, PCB, Coplnar PCB and eggshell parameters for reproduction in the white-tailed sea eagle (*Haliaeetus albicilla*) in Sweden. *Ambio* **31**(5).
HELCOM (1998) Agenda 21 for the Baltic Sea region: sustainable development of the agricultural sector in the Baltic Sea region. *Baltic Sea Environmental Proceedings* **74**.
Henderson S (2001) Personal communication, monitoring data from the former East of Scotland Water.
Henriques W, Jeffers RD, Lacher TE, and Kendall RJ (1997) Agrochemical use on banana plantations in Latin America: perspectives on ecological risk. *Environ Toxicol Chem* **16(**1): 91–100.

Herricks EE (ed.) (1995) *Stormwater Runoff and Receiving Systems: Impact, Monitoring and Assessment.* Lewis Publishers, New York, Tokyo & London.

Higginbotham S, Jones RL, Gatzweiler E and Mason PJ (1999) Point source pesticide contamination: quantification and practical solutions. *Proceedings 1999 Brighton Conference on Weeds,* British Crop Protection Council, Farnham

Holdgate MW (1970) *The Seabird Wreck of 1969 in the Irish Sea.* A Report of the National Environment Research Council, UK.

Hollemans WA (2002) Weed control without herbicides: different methods in three provinces. In *Proceedings of the 6th International Conference on Diffuse Pollution*, Amsterdam, 30 September–4 October. International Water Association, in co-operation with the Netherlands Association on Water Management (NVA) and Aquatech, Amsterdam.

Horner RR, Skupien JJ, Livingston EH and Shaver HE (1994) *Fundamentals of Urban Runoff Management: Technical and Institutional Issues.* Terrene Institute in cooperation with US Environmental Protection Agency, Washington, DC.

Hydrologic Engineering Centre (1976) *Storage, Treatment, Overflow Runoff Model 'STORM'.* US Army Corps of Engineers, Davis, CA.

IAWQ (1996) *Nature's Way.* Video. IWA Publishing, London.

Ichiki A and Yamada K (1999) Study of characteristics of pollutant runoff into Lake Biwa, Japan. *Water Science and Technology* **39**(12): 17–25.

IEH (1999) *IEH Assessment on the Ecological Significance of Endocrine Disruption: Effects on Reproductive Function and Consequences for Natural Populations.* Assessment A4. MRC Institute for Environment and Health, Leicester.

Iwugo K, Andoh R, Feest A (2001) Cost-effective urban drainage and wastewater management systems. *Water 21*, April 2001, 51–4. IWA Publishing, London.

Iwugo K, D'Arcy BJ, Heath RGM and Andoh R (2003) Breaking the pollution cycle in poor African settlements. Paper presented at 29th WEDC Conference, 22–26 September, Abuja, Nigeria.

Jackson JE (1999) The Tualatin River: a water quality challenge. In AC Rowney, P Stahre and LA Roesner (eds) *Sustaining Urban Water Resources in the 21st Century.* ASCE, Reston, VA.

Jefferies C (2003) *SUDS in Scotland – The Monitoring Programme: Final report of the Scottish Universities SUDS Monitoring Group.* SNIFFER, Edinburgh.

Jenkins A, Ellis JB and Ferrier RC (2000) Modelling diffuse pollution. In BJ D'Arcy, JB Ellis, RC Ferrier, A Jenkins and R Dils (2000) *Diffuse Pollution Impacts: the Environmental and Economic Impacts of Diffuse Pollution in the UK.* Chartered Institute of Water and Environmental Management (CIWEM) and Terence Dalton Publishers, Lavenham.

Johnes PJ (1996) Nutrient loss coefficients, based on studies of arable fields in Windrush Catchment. *Journal of Hydrology* **183**: 323–49.

Joly C (1993) Plant nutrient management and environment. In *Prevention of Water Pollution by Agriculture and Related Activities.* Proceedings of the FAO Expert Consultation, Santiago, Chile, 20–23 October. Water Report No. 1. FAO, Rome.

Kannan K, Blankenship AL, Jones PD and Geisy JP (2000) Toxicity reference values for the toxic effects of polychlorinated biphenyls to aquatic mammals. *Hum. Ecol. Risk Assess.* **6**: 181–201.

Kay D, Wither AW and Jenkins A (2000) Faecal pathogens. In D'Arcy BJ, Ellis JB, Ferrier RC, Jenkins A and Dils R (eds) *Diffuse Pollution Impacts: The Environmental and Economic Impacts of Diffuse Pollution in the UK*. Chartered Institute of Water and Environmental Management (CIWEM), Terence Dalton Publishers, Lavenham.

Kim L-H, Kayhanian M and Stenstrom MK (2002) Prediction of event mean concentrations and first flush effects using a mass interpolation wash-off model for highway runoff. In *Proceedings of the 6th International Conference on Diffuse Pollution*, Amsterdam, 30 September–4 October. International Water Association, in co-operation with the Netherlands Association on Water Management (NVA) and Aquatech, Amsterdam.

Knight RL (1997) Wildlife habitat and public use benefits of treatment wetlands. *Water Science and Techology* **35**(5): 35–43.

Kreitler CW and Jones DC (1975) Natural soil nitrate: the cause of the nitrate contamination in groundwater in Runnels County, Texas. *Groundwater* **15:** 160–9.

Kulzer L (1989) *Considerations for the Use of Wet Ponds for Water Quality Enhancement*. Munici. Metro, Seattle, WA.

Kumm, KI. (1990) Incentive policies in Sweden. In J Baden and SB Lovejoy (eds) *Agriculture and Water Quality: International Perspectives*. Lynne Riemer, London.

Lambin X, Telfer S, Aars J, Denny R and Griffin CY (2000) Metapopulation persistence of water voles in face of mink predation. In Keeble and Grey (eds) *Proceedings of the Water Vole Conference* 16 October 1999. People's Trust for Endangered Species, London.

Langton T, Beckett C and Foster J (2001) *Great Crested Newt Conservation Handbook*. Froglife, Halesworth.

Lann H (2001) Personal communication, based on presentation at EU workshop in Brussels, 12 October 2001

Larm J (1999) A case study of the stormwater treatment facilities in Flemingsbergsviken, south of Stockholm. In AC Rowney, P Stahre and LA Roesner (eds) *Sustaining Urban Water Resources in the 21st Century.* Proceedings of a conference in Malmö, Sweden, 7–12 September 1997. ASCE, Reston, VA.

Larsson T and Karppa P (1999) Integrated stormwater management in the city of Malmö, Sweden. In AC Rowney, P Stahre and LA Roesner (eds) *Sustaining Urban Water Resources in the 21st Century.* Proceedings of a conference in Malmö, Sweden, 7–12 September 1997. ASCE, Reston, VA.

Le Gallic H-C (2001a) Personal communication: Brittany and the restoration of its water quality. Bretagne Eau Pure, Rennes.

Le Gallic H-C (2001b) Brittany and the restoration of its water quality. Presentation at the Diffuse Pollution Workshop, 12 October, Directorate General Environment, European Commission, Brussels.

Li Y (1999) Agricultural diffuse pollution from fertilizers and pesticides in China. Paper presented at the International Workshop, Teolo, Italy, *Water Science and Technology,* **39**(3).

Lindsay and Doll (1998) Financing retro-fit projects: the role of stormwater utilities. EPA National Conference on Retro-fit Opportunities for Water Resource Protection in Urban Environments. 9–12 February. Chicago, IL.

Livingston EH (1995) Lessons learned from a decade of stormwater treatment in Florida. In EE Herricks (ed.) *Stormwater Runoff and Receiving Systems: Impact, Monitoring and Assessment*. CRC Press, Inc. New York.

Livingston EH, Shaver E and Skupien JJ (1997) *Operation, Maintenance, and Management of Stormwater Management Systems*. Watershed Management Institute, Inc., in co-operation with US Environmental Protection Agency, Washington, DC.

Loehr, RC (1972) Agricultural runoff: characterization and control. *J. Sanitary Eng. Div.,* **98**: 923–99.

Lord EI and Shepherd MA 1996. Effects of long-term straw incorporation and drilling date on nitrate leaching: lysimeter study. *Aspects of Applied Biology* **47**: 417–20.

Macdonald KB (2003) The Effectiveness of Certain Sustainable Urban drainage Systems in controlling Flooding and Pollution from Urban Runoff. PhD thesis, University of Abertay Dundee.

Macdonald KB and Jefferies C (2001) Performance comparison of porous paved and traditional car parks. In *Proceedings of 1st National Conference on Sustainable Drainage*, June 2001, Coventry.

Macdonald KB and Jefferies C (2003) Performance comparison of two swales. In CJ Pratt, JW Davies, AP Newman and JL Perry (eds) *Proceedings of 1st National Conference on Sustainable Drainage,* June 2001, Coventry.

MacDonald K and Williams E (2001) Legislation, regulation and industrial competitiveness: a comparative study of regulation in Denmark, Portugal and Scotland. Paper presented at the Environmental Policy and the Costs of Compliance Workshop, September, London School of Economics, London.

MacDonald RW, Barrie LA, Bidleman TF, Diamond ML Gregor DJ, Semkin RG and Strachan WMJ (2000) Contaminants in the Canadian Arctic: five years of progress in understanding sources, occurrence and pathways. *Science of the Total Environment* **254**(2–3): 93–235.

Maitland PS and Campbell RN (1992) *Freshwater Fishes of the British Isles*. The New Naturalist Lbirary. HarperCollins Publishers London.

Marsalek, J and Torno HC (1993) *Proceedings 6th International Conference on Urban Storm Drainage.* Seapoint Publishing, Victoria, BC.

Marsalek J, White E, Zemen E and Sieker H (2001) *Advances in Urban and Agricultural Runoff Source Controls*. Kluwer Academic, Dordrecht.

Marshall D (2002) Fraser River Basin, Canada: sustainability applied to catchment and watershed management. International Water Association Leading Edge Conference 'Sustainability in the Water Sector', 25–26 November, Venice. (Abstracts and CD of presentations from IWA, London.)

Marttinen SK (2002) Treatment of pthalate load from non-point and point sources at a sewage treatment plant. In *Proceedings of the 6th International Conference on Diffuse Pollution*, Amsterdam, 30 September–4 October. International Water Association, in co-operation with the Netherlands Association on Water Management (NVA) and Aquatech, Amsterdam.

Mason CF and MacDonald SM (1993) Impact of organochloride pesticide residues and PCBs on otters (*Lutra lutra*): a study from Western Britain. *Science of the Total Environment* **138(**1/3): 127–45.

Mawdsley JL, Bargett RD, Merry RJ, Pain B and Theodorou MK (1995) Pathogens in livestock waste: their potential for movement through soil and environmental pollution. *Applied Soil Ecology,* **2**: 1–5.

McKissock G (2002) persoanl communication: unpublished deliberations of the Sustainable Urban Drainage Welsh Working Party.

McKissock G, D'Arcy BJ and Jeffries C (2001) Sustainable urban drainage: a case study. In *Innovative technologies in urban drainage*, Novatech 4th International Conference, 25–27 June, Lyon. Novatech, Lyon.

McKissock G, D'Arcy BJ, Wild TC, Usman F. and Wright PW (2003a) An evaluation of SUDS guidance in Scotland. *Diffuse Pollution & Basin Management*, Proceedings of the 7th IWA International Diffuse Pollution Conference, 17–22 August, Dublin. International Water Association, London.

McKissock K, McKissock G, Barbarito B and D'Arcy BJ (2003b) A methodology for retrofitting SUDS technology in existing urban areas. Unpublished SEPA technical report, DPI No. 21 / BJD / 1st draft.

McNeil A (1998) Best management practices and contingency planning for spillage emergencies. Clean Stream Conference *Surveyor* 25 June 1998.

McNeil A (2000) Natural Remedy *Surveyor* 20 April 2000.

McNeil A (2003) personal communication (SEPA action plan project in preparation).

Meals DW and Hopkins RB (2001) Phosphorus reductions following riparian restoration in two agricultural watersheds in Vermont, USA. *Water Science and Technology* **45**(9): 51–60.

Melching CS and Avery CC (1990) *An Introduction to Watershed Management for Hydrologists.* V.U.B. Hydrologie, Free University of Brussels, Brussels

Mitchell G (2001) *The Quality of Urban Stormwater in Britain and Europe: Database and Recommended Values for Strategic Planning Models.* School of Geography, The University of Leeds, Leeds.

Mohaupt V, Bach M, and Behrendt H (2000) Overview on diffuse sources of nutrients, pesticides and heavy metals in Germany: methods, results and recommendations for water protection policy. In *Proceedings 4th International Conference on Diffuse Pollution.* 16–21 January, Bangkok. *Water Science and Technology* **44**(7).

Moir SE, Svoboda I, Sym G and Clark J (2003) The treatment of dairy washings from milking equipment using a sequencing batch reactor and a reedbed treatment system. In I McTaggart and L Gairns (eds) *Agriculture, Waste and the Environment.* SAC/SEPA Biennial Conference. 26–28 March, Edinburgh. SAC, Edinburgh.

Moldenhauer W, Kemper W and Langdale G (1994) Long-term effects of tillage and crop residue management. In G Langdale and W Moldenhauer (eds) *Crop Residue Management to Reduce Erosion and Improve Soil Quality-Southeast.* CR39. Agricultural Research Service, USDA, Washington, DC.

Moore IC, Madison FW, and Schneider RR (1979) Estimating phosphorus loading from livestock wastes: some Wisconsin results. In RC Loehr (ed.) *Best Management Practices for Agriculture and Silviculture.* Van Nostrand-Reinhold, New York.

Moore PA, Daniel TC and Edwards DR (1998) Reducing phosphorus runoff and inhibiting ammonia loss from poultry manure with aluminium sulphate. *Proceedings of the OECD Workshop on Practical and Innovative Measures for the Control of Agricultural Phosphorus Losses to Water*, 16–19 June, Antrim, Northern Ireland .

Murray JE and Cave KA (1999) The Rouge Project: building institutional and regulatory frameworks necessary for watershed approaches to wet weather pollution management. In AC Rowney, P Stahre and LA Roesner (eds) *Sustaining Urban Water Resources in the 21st Century.* Proceedings of a conference in Malmö, Sweden, 7–12 September 1997. ASCE, Reston, VA.

Newman AP, Shuttleworth A, Puehmeier T, Wing Ki K and Pratt CJ (2003) Recent developments in oil retaining porous pavements. *Proceedings Second National conference on Sustainable Drainage* 23–24 June, Coventry University.

Nicholson RJ and Baldwin DJ (1999) Opportunities for waste minimisation in agriculture and production of a self-help manual for farmers. SEPA/SAC Bi-Annual Conference, March, Edinburgh University.

Nisbet TR (1994) Forests and water guidelines: how well do they tackle the water issues? In IR Brown (ed.) *Forests and Water* Proceedings of a discussion meeting, 25–27 March, Heriot-Watt University, Edinburgh.

Nisbet TR, Welch D and Doughty R (1998) The role of forest management in controlling diffuse pollution. In T Petchey, B D'Arcy and A Frost (eds) *Diffuse Pollution and Agriculture.* The Scottish Agricultural College, Aberdeen.

Novotny V (ed) (1995) *Non-point Pollution and Urban Stormwater Management.* TECHNOMIC Publishing Co., Lancaster, PA.

Novotny V (1996) Integrated water quality management. *Water Science and Technology* **33**(4–5): 1–7.

Novotny V (1999) Diffuse pollution from agriculture: a worldwide outlook. *Water Science and Technology* **39**(3): 1–13.

Novotny V (2000) Root causes of diffuse pollution: developing countries. *Proceedings 4th International Conference on Diffuse Pollution,* 17–21 January, Bangkok. *Water Science and Technology* **44**(7).

Novotny V (2003) *Water Quality: Diffuse Pollution and Watershed Management.* 2nd edn. John Wiley & Sons, New York.

Novotny V and D'Arcy BJ (eds) (1999) *Diffuse Pollution '98: Proceedings of the Edinburgh IAWQ 3rd International Conference on Diffuse Pollution.* 21 August–4 September 1998, Edinburgh. Special issue of *Water, Science and Technology* **39(**12).

Novotny V and Olem H (1994) *Water Quality: Prevention Identification and Management of Diffuse Pollution.* Van Nostrand Reinhold, New York, reprinted and distributed by J. Wiley & Sons, New York.

Novotny V *et al.* (1997) *A Comprehensive UAA Technical Reference.* Water Environment Research Foundation, Alexandria, VA.

Novotny, V., D.W. Smith, D.A. Kuemmel, J. Mastriano, and A. Bartošová (1999) *Urrban and Highway Snowmelt: Minimizing the Impact on Receiving Water.* Water Environment Research Foundation, Alexandria, VA.

Novotny V, Booth D, Clark D and Griffin R (2000) Risk-based urban watershed management under conflicting objectives. *Water Science and Technology* **43**(5): 69–78.

Olem, H (ed.) (1993) *Diffuse Pollution.* Special issue of *Water Science amd Technology* **29**(3–4).

Odell B (2001) Uncertain world: report on future worlds – environmental changes and their effects. *Water and Environment Manager* **6**(4): 6–7.

OECD (1994) *Managing the Environment: The Role of Economic Instruments*. OECD, Paris.

O'Keefe B, D'Arcy BJ, Davidson J, Babarito B and Clelland B (2005) Urban diffuse sources of faecal indicators. *Proceedings of the 7th International Specialist Conference on Diffuse Pollution and Basin Management.* Vol. 3. 17–22 August, *Water Science and Technology* **51**(3).

Omernik JM and Gallant AL (1990) Defining ecoregions for evaluating environmental resources. In *Global Natural Resource Monitoring and Assessment: Preparing for the 21st Century. Proceedings of the International Conference and Workshop*, Vol. 2. American Society for Photogrammetry and Remote Sensing, Bethesda, MD.

Ongley ED (1996) *Control of Water Pollution from Agriculture*. FAO, Rome.

Ontario Ministry of Environment (1993) *Guidelines for the Protection and Management of Aquatic Sediment Quality in Ontario*. Ontario Ministry of the Environment.

O'Riordan T (1995) *Environmental Science for Environmental Management.* Prentice Hall, London.

Parr W, Andrews K, Mainstone CP and Clarke SJ (1999) *Diffuse Pollution: Sources of N and P*. WRC report No. DETR 4755 to the Department for the Environment Transport and the Regions. Water Research Centre.

Patty L, Gril JJ, Real B and Guyot C (1995) Grassed buffer strips to reduce herbicide concentration in run-off: preliminary study in Western France. In A Walker (ed.) *Pesticide Movement to Water.* BCPC Monograph No 62. BCPC Publications, Alton, UK.

Pratt CJ (1999a) Use of permeable, reservoir pavement constructions for stormwater treament and storage for re-use. *Water Science and Techology* **39**(5): 145–51.

Pratt CJ (1999b) Design guidelines for porous/permeable pavements. In AC Rowney, P Stahre and LA Roesner (eds) *Sustaining Urban Water Resources in the 21st Century.* Proceedings of a conference in Malmö, Sweden, 7–12 September 1997. ASCE, Reston, VA.

Pratt CJ, Newman AP and Bond PC (1999) Mineral oil biodegradation within a permeable pavement: long-term observations. *Water Science and Technology* **39(**2): 103–9.

Pretty JN, Brett C, Gee D, Hine RE, Mason CF, Morrison JIL, Raven H, Rayment MD and Bilj G van der (2000) An assessment of the total external costs of UK agriculture. *Agricultural Systems*. (In press).

Puckett LJ (2004) Hydrogeologic controls on the transport and fate of nitrate in groundwater beneath riparian buffer zones: results from studies across the United States. In *Proceedings of the 6th International Conference on Diffuse Pollution*, Amsterdam, 30 September–4 October. International Water Association, in co-operation with the Netherlands Association on Water Management (NVA) and Aquatech, Amsterdam.

Raimbault G (1999) French developments in reservoir structures. In AC Rowney, P Stahre and LA Roesner (eds) *Sustaining Urban Water Resources in the 21st Century.* Proceedings of a conference in Malmö, Sweden, 7–12 September 1997. ASCE, Reston, VA.

Richards T (1999) Restoring urban streams in Atlanta: Metro Atlanta urban watersheds initiative. In AC Rowney, P Stahre and LA Roesner (eds) *Sustaining Urban Water Resources in the 21st Century.* Proceedings of a conference in Malmö, Sweden, 7–12 September 1997. ASCE, Reston, VA.

Robins JWD (1985) Best management practices for animal production. *Proceedings Non-point Pollution Symposium,* Marquette University, Milwaukee, WI.

Robinson R (2000) The environment protection viewpoint. In AM Petchey, BJ D'Arcy and CA Frost (eds) *Agriculture and Waste: Management for a Sustainable Future. Proceedings of a Joint SAC/SEPA Conference*, 31 March–2 April 1997, Edinburgh. Scottish Agricultural College, Aberdeen.

Roesner LA (1999a) The hydrology of urban runoff quality management. In AC Rowney, P Stahre and LA Roesner (eds) *Sustaining Urban Water Resources in the 21st Century.* Proceedings of a conference in Malmö, Sweden, 7–12 September 1997. ASCE, Reston, VA.

Roesner LA (1999b) Urban runoff pollution: summary thoughts. the state of practice today and for the 21st century. *Water Science and Technology* **39**(12): 353–60.

Roesner LA, Howard RM, Mack BW, and Ramdett CA (1999) Integrating stormwater management into urban planning in Orlando, Florida. In AC Rowney, P Stahre and LA Roesner (eds) *Sustaining Urban Water Resources in the 21st Century.* Proceedings of a conference in Malmö, Sweden, 7–12 September 1997. ASCE, Reston,VA.

Roesner LA, Campbell N and D'Arcy BJ (2001) Master planning stormwater management functions for the Dunfermlin, Scotland Expansion Project. *Innnovative Technologies in Urban Drainage* Vol. 1. Novatech, Lyons.

ROSPA (1999) *Safety at Inland Water Sites: Operational Guidelines.* The Royal Society for the Prevention of Accidents, in association with The Royal Life-saving Society.

Royal Commission on Environmental Pollution (1996) *Sustainable Use of Soil.* HMSO, London.

Rowney AC, Stahre P and Roesner LA (eds) (1999) *Sustaining Urban Water Resources in the 21st Century.* Proceedings of a conference in Malmö, Sweden, 7–12 September 1997. ASCE, Reston, VA.

RSPB (1996) *Crisis in the Hills: Overgrazing in the Uplands.* Royal Society for the Protection of Birds, Sandy.

Runolfsson S (1998) Soil erosion and conservation in iceland. In T Petchey, B D'Arcy and A Frost (eds) *Diffuse Pollution and Agriculture II. Proceedings of a Joint SAC/SEPA Conference*, 9–11 April, 1997, Edinburgh. The Scottish Agricultural College, Aberdeen.

Sanchez PA (1992) Tropical region soils management. In VW Ruttan (ed.) *Sustainable Agriculture and the Environment.* Westview Press, Boulder, CO.

Sansom AL (1999a) *Farmer/Regulator Relationships and How to Improve Them.* Nuffield Farming Scholarship, Yorkshire Agricultural Society Award. Nuffield Farming Scholarships Trust, Market Harborough, UK.

Sansom, AL (1999b) Upland vegetation management: the impacts of overstocking. *Water Science and Techology* **39(**12): 85–92.

Schueler TR (1987) *Controlling Urban Runoff: A Practical Manual for Planning and Designing Urban BMPs.* Metropolitan Council of Governments, Washington, DC.

Schueler TR (1995) *Site Planning for Urban Stream Protection.* Centre for Watershed Protection and Metropolitan Washington Council of Governments, Washington, DC.

Schueler TR (2000) Comparative pollutant removal capability of stormwater treatment practices. Article 64, Technical Note #95, *Watershed Protection Techniques* **2**(4): 515–20.

Schueler TR and Galli J (1995) The environmental impact of stormwater ponds. In EE Herricks (ed.) *Stormwater Runoff and Receiving Systems: Impact, Monitoring and Assessment.* Lewis Publishers, New York, Tokyo & London.

Schueler TR, Kumble PA and Heraty MA (1992) *A Current Assessment of Urban Best Management Practices*. Metropolitan Washington Council of Governments, Washington, DC.

Scottish Executive (2002) *Custodians of Change*. Report of the Agriculture and Environment Working Group. The Stationery Office, Edinburgh.

SEPA (1999) *Improving Scotland's Water Environment*. SEPA State of the Environment Report. Scottish Environment Protection Agency, Stirling.

SEPA (2000a) *Ponds, Pools and Lochans: Guidance on Good Practice in the Management and Creation of Small Waterbodies in Scotland*. Scottish Environment Protection Agency, Stirling.

SEPA (2000b) *Watercourses in the Community: A Guide to Sustainable Watercourse Management in the Urban Environment*. Scottish Environment Protection Agency, Stirling.

SEPA (2002) *The Annan Catchment Coordination Project*. Scottish Environment Protection Agency, Stirling.

SEPA (2003) *Protecting and Improving the Environment Through Regulation, SEPA's Vision for Regulation*. And *Consultation on the Outputs of the Effective Regulation programme – September 2003*. Scottish Environment Protection Agency, Stirling.

Sfriso A, Pavoni B, Marcomini A, Orio AA (1988) Annual variations of nutrients in the lagoon of Venice. *Marine Pollution Bulletin* **19**(2): 54–60.

Sheail, J (1985) *Pesticides and Nature Conservation: The British Experience, 1950–1975*. Clarendon Press. Oxford.

Skinner JA, Lewis KA, Bardon KS, Tucker P, Catt JA and Chambers BJ (1997) An overview of the environmental impact of agriculture in the UK. *Journal of Environmental Management* **50:** 11–128.

Smith KA, Chalmers AG, Chambers BJ and Christie P (1998) Organic manure phosphorus accumulation, mobility and management. *Soil Use and Management*, **14**: 154–9.

Smulders L (2002) Prevention of diffuse water pollution due to leaching metals in building construction. In *Proceedings of the 6th International Conference on Diffuse Pollution*, Amsterdam, 30 September–4 October. International Water Association, in co-operation with the Netherlands Association on Water Management (NVA) and Aquatech, Amsterdam.

South African Government (1998) The National Water Act (Act 36 of 1998).

Spitzer A (2003) personal communication. WWTC, Abertay University, Dundee, UK.

Spray C (2002) The water industry vision. Presentation for Joint Water UK/Royal Society for the Protection of Birds Workshop: *Diffuse Pollution, Floods and Farming*, 22 April, Commonwealth Institute, London. Water UK, Information & Learning, London.

Spruill TB (2004) Effectiveness of riparian buffers in controlling discharge of nitrate to streams in selected hydrogeologic settings of the North Carolina coastal plain. In *Proceedings of the 6th International Conference on Diffuse Pollution*, Amsterdam, 30 September–4 October. International Water Association, in co-operation with the Netherlands Association on Water Management (NVA) and Aquatech, Amsterdam.

Stahre P (1996) Quotes from the video *Nature's Way*. IWA Publishing, London.

Stewart BA, Woolhiser DA, Wischmeier WH, Caro JH and Frere MH (1975) In *Control of Water Pollution from Cropland. Volume 1. A Manual for Guideline Development*. US

Department of Agriculture report number ARS-H-5-1. US Government Printing Office, Washington, DC.

Stovin VR and Swan AD (2003) Application of a retrofit SUDs decision-support framework to a UK catchment. *Proceedings Second National Conference on Sustainable Drainage,* 23–24 June, Coventry University, Coventry, UK.

Strachan R (1998) *Water Vole Conservation Handbook*. English Nature, the Environment Agency, and Wildlife Conservation Research Unit (University of Oxford), Bristol.

Straskraba M. (1996) Ecotechnological methods for managing non-point source pollution in watersheds, lakes and reservoirs. *Water Science and Technology* **33(**4–5).

Strecker E, Quigley M, Urbonas B, Jones J and Clary J (2001) Determining urban stormwater BMP effectiveness. *Journal of Water Resources Planning and Management*, **127**(3): 143–201.

Swedish Environmental Protection Agency (1997) *Environmental Taxes in Sweden: Economic Instruments of Environmental Policy*. Swedish Environmental Protection Agency, Stockholm.

Thompson P (1989) *Poison Runoff: A Guide to State and Local Control of Non-point Source Water Pollution.* Natural Resources Defense Council, New York.

Thomsen IK and Christensen BT (1999) Nitrogen conserving potential of successive ryegrass catch crops in continuous spring barley. *Soil Use and Management* **15**: 195-200.

Thorarinsson S (1974) Thjodin lividi en skogurinn do (The nation lived but the forest died). *Arsrit Skograktarfelags Islands* (*Yearbook of the Forest Society of Iceland*). Forest Society of Iceland, Reykjavik.

Tilborg WJM van, Frankena RB and Bosch J van den (2002) Views related to ongoing developments around zinc, copper and lead in the building and construction industry. In *Proceedings of the 6th International Conference on Diffuse Pollution*, Amsterdam, 30 September–4 October. International Water Association, in co-operation with the Netherlands Association on Water Management (NVA) and Aquatech, Amsterdam.

Tonmanee N and Kanchanakool N (1998) Problem-solving on diffuse (non-point) pollution in Thailand. In V Novotny and BJ D'Arcy (eds) *Diffuse Pollution '98: Proceedings of the Edinburgh IAWQ 3rd International Conference on Diffuse Pollution.* 21 August–4 September 1998, Edinburgh. Scottish Environmental Protection Agency, Edinburgh.

Torstensson T and de Pilar Castillo M (1997) Use of biobeds in Sweden to minimise environmental spillages from agricultural spraying equipment. *Pesticide Outlook* June, 24–7.

Tuininga K (2002) Voluntary and enforcement action to reduce livestock impacts on Canadian watercourses. In *Proceedings of the 6th International Conference on Diffuse Pollution*, Amsterdam, 30 September–4 October. International Water Association, in co-operation with the Netherlands Association on Water Management (NVA) and Aquatech, Amsterdam.

Tyler CR (1998) Endocrine disruption in wildlife: a critical review of the evidence. *Critical Reviews in Toxicology* **28**(4): 319–62.

Uchimura K, Nakamura E and Fujita S (1997) Characteristics of stormwater runoff and its control in Japan. *Water Science and Technology* **36**(8–9): 141–7.

UNEP (1997) Report of the First Meeting of the Steering Group for the Global International Waters Assessment. 24–27 February, Geneva.

UNIDO (1984) Global Overview of Pesticide Sub-sector. Unpublished sectoral working paper. United Nations Industrial Development Organization. Vienna.

Urbonas B (1999a) Design and selection guidance for structured BMPS. In AC Rowney, P Stahre and LA Roesner (eds) *Sustaining Urban Water Resources in the 21st Century.* Proceedings of a conference in Malmö, Sweden, 7–12 September 1997. ASCE, Reston,VA.

Urbonas, B (1999b) *Urban Storm Drainage: Criteria Manual.* Vol. 3 *Best Management Practices*. Urban Drainage and Flood Control District, Denver, CO.

Urbonas, B and Stahre P (1993) *Best Management Practices and Detention for Water Quality, Drainage and CSO Management.* Prentice-Hall, Englewood Cliffs, NJ.

Urbonas B, Roesner LA and Guo CY (1996) Hydrology for optimal sizing of urban run-off treatment control systems. *Water Quality International* January/February: 30–3.

US Congress, Office of Technology Assessment (1990) Technologies to improve nutrient and pest management. In *Beneath the Bottom Line: Agricultural Approaches to Reduce Agrichemical Contamination of Groundwater*. Report OTA-F-418. U.S. Government Printing Office, Washington, DC.

US Department of Agriculture (1994) *Evaluating the Effectiveness of Forestry Best Management Practices in Meeting Water Quality Goals or Standards.* Miscellaneous publication 1520. US Department of Agriculture, Forest Service, Washington, DC.

USEPA (1983) *Results of the Nationwide Urban Runoff Program. Vol. 1, Final Report.* Water Planning Division, Washington, DC.

USEPA (1986) *Methodology for Analysis of Detention Basins for Control of Urban Runoff Quality*. EPA 440/5-87-0001. USEPA, Washington, DC.

USEPA (1993) *Guidance Specifying Management Measures For Sources of Non-point Pollution In Coastal Waters.* United States Environmental Protection Agency, Office of Water, Washington, DC.

USEPA (1997a) *Monitoring Guidance for Determining the Effectiveness of Non-point Source Controls*. US Environmental Protection Agency, Non-point Source Branch, Office of Water, Washington, DC.

USEPA (1997b) Sny Magill Creek: the new standard agricultural practices. In *Success Stories: Volume II, Highlights of State and Tribal Non-point Source Programs.* Environmental Protection Agency, Washington, DC.

USEPA (1998) *National Water Quality Inventory: 1996 Report to Congress,* Environmental Protection Agency, Washington, DC.

USEPA (2000) *US National Water Quality Inventory, Section 305(b) Report,* Environmental Protection Agency, Washington, DC.

USGS (2003) Monitoring the effectiveness of urban best management practices in improving water quality of Englesby Brook, Burlington, Vermont. USGS Fact Sheet 114-00. Available online http://water.usgs.gov/pubs/fs/FS-114-00/

Usman F (2000) Effective Pollution Prevention Campaigns. Unpublished joint technical report, Scottish Environment Protection Agency and the Environment Agency, Bristol.

Usman F, D'Arcy BJ, Richardson H and Chatfield P (1998) The Prevention of Oil and Chemical Pollution in the UK. Unpublished SEPA technical report. Scottish Environment Protection Agency, Edinburgh.

Usman F, Fernandes T, Fernie J, Read P and Hundal J (2000) Prevention of oil and chemical pollution: proactive initiatives in the UK and Canada. *Journal of the Chartered Institute of Water and Environmental Management.*

Vinten AJA, Davies R, Castle K and Baggs EM (1998) Control of nitrate leaching from a nitrate vulnerable zone using paper mill waste. *Soil Use and Management* **14**: 44–51.

Walker, WW (1987) Phosphorus removal by urban runoff detention basins. *Lake Reservoir Management* **3**: 314–26.

Walker K, D'Arcy BJ, McKissock G and Wilby N (2000) SUDS: The Scottish Experience. Paper presented at 'British Pond Landscapes: Moving Towards Sustainability' Conference, 18–19 September, University College, Chester.

Wanielista M., Yousef YA, and Taylor JS (1981) *Stormwater Management to Improve Lake Water Quality*. EPA 600/12-82-084. US Environmental Protection Agency, Washington, DC.

Watershed Information Network News (1998) French Creek Watershed Advisory Group. In *Runoff Report, National Forum on Non-point Source Pollution's Demonstration Projects*. Terrene Institute, Reston, VA.

Watt, WE, Marsalek, J and Anderson BC (1999) Stormwater pond perceptions vs. realities: a case study. In AC Rowney, P Stahre and LA Roesner (eds) *Sustaining Urban Water Resources in the 21st Century.* Proceedings of a conference in Malmö, Sweden, 7–12 September 1997. ASCE, Reston,VA.

Weitman D (1996) Controlling diffuse pollution by best management practices. In T Petchey, B D'Arcy and A Frost (eds) *Diffuse Pollution and Agriculture* The Scottish Agricultural College, Aberdeen.

Wild TC, Jefferies C and D'Arcy BJ (2002) *SUDS in Scotland: The Scottish SUDS Database.* SNIFFER Report SR(02)09. Scotland and Northern Ireland Forum for Environmental Research, Edinburgh.

Williams M (2002) Natural capital: the account we all continue to plunder. Keynote presentation at International Water Association Leading Edge Conference 'Sustainability in the Water Sector', 25–26 November, Venice. (Abstracts and CD of presentations from IWA, London.)

Williams E, MacDonald K and Kind V (2002) Unravelling the competitiveness debate. *European Environment* **12**: 284–90.

Wilson C and Clarke R (2002) Persistent Pollutants in Freshwater Sediments. Unpublished SEPA report. Edited version available as DPI Report No. 7, from SEPA Diffuse Pollution Initiative, Scottish Environment Protection Agency, Edinburgh.

Wilson C, Clarke R, D'Arcy BJ, Heal KV and Wright PW (2003) Persistent Pollutants Urban Rivers Sediments Survey: implications for pollution control. *Proceedings of the 7th International Conference on Diffuse Pollution*, August, Dublin. International Water Association, London.

Wise SW and Hann MJ (2000) An holistic approach to the management of wastewater sludge phosphorus recycled to agricultural land. In AM Petchey, BJ D'Arcy and CA Frost (eds) *Agriculture and Waste: Management for a Sustainable Future. Proceedings of a Joint SAC/SEPA Conference*, 31 March–2 April 1997, Edinburgh. Scottish Agricultural College, Aberdeen.

Withers PJA and Sylvester-Bradley R, 1999. Nitrogen fertiliser requirements of cereals following grass. *Soil Use and Management* **15**: 221–9.

Worrall P, Peberdy KJ and Millett MC (1997) Constructed wetlands and nature conservation. *Water Science and Techoolgy* **35(**5): 205–13.

Wyer MD, O'Neill JG, Goodwin V, Kay D, Jackson GF, Tanguy L and Briggs J (1997) Non-sewage derived sources of faecal indicator organisms in coastal waters: case studies from Yorkshire and Jersey, Great Britain. In D Kay and C Fricker (eds) *E. coli: Problem or Solution*. Royal Society of Chemistry, London.

Yamada K, Nishikawa K and Sugihara M (2002) Study on removal by soil of pollutants discharged from road surface during storm events. In *Proceedings of the 6th International Conference on Diffuse Pollution*, Amsterdam, 30 September–4 October. International Water Association, in co-operation with the Netherlands Association on Water Management (NVA) and Aquatech, Amsterdam.

Younger P (2000) Iron. In BJ D'Arcy, JB Ellis, RC Ferrier, A Jenkins and R Dils (eds) *Diffuse Pollution Impacts: The Environmental and Economic Impacts of Diffuse Pollution in the UK*. CIWEM, London.

Zandbergen P (2002) Effectiveness of public education as an urban stormwater BMP. In *Proceedings of the 6th International Conference on Diffuse Pollution*, Amsterdam, 30 September–4 October. International Water Association, in co-operation with the Netherlands Association on Water Management (NVA) and Aquatech, Amsterdam.

Zeegers I, Klavers H, Bijsterbosch J and Blecourt C de (2002) Enforcement reduces diffuse pollution. In *Proceedings of the 6th International Conference on Diffuse Pollution*, Amsterdam, 30 September–4 October. International Water Association, in co-operation with the Netherlands Association on Water Management (NVA) and Aquatech, Amsterdam.

Index